D0008195

INTERMEDIATE CHEMISTRY DIVISION

General Editor: J. E. SPICE

CHEMICAL KINETICS

AND

SURFACE AND COLLOID CHEMISTRY

CHEMICAL KINETICS

by

A. F. TROTMAN-DICKENSON

Professor of Chemistry
University College of Wales, Aberystwyth

SURFACE AND COLLOID CHEMISTRY

by

G. D. PARFITT

Lecturer in Physical Chemistry
University of Nottingham

PERGAMON PRESS

OXFORD · LONDON · EDINBURGH · NEW YORK
TORONTO · PARIS · FRANKFURT

Pergamon Press Ltd., Headington Hill Hall, Oxford
4 & 5 Fitzroy Square, London W.1
Pergamon Press (Scotland) Ltd., 2 & 3 Teviot Place, Edinburgh 1
Pergamon Press Inc., 44–01 21st Street, Long Island City, New York 11101
Pergamon of Canada Ltd., 6 Adelaide Street East, Toronto, Ontario
Pergamon Press S.A.R.L., 24 rue des Écoles, Paris 5e
Pergamon Press GmbH, Kaiserstrasse 75, Frankfurt-am-Main

First edition 1966
Library of Congress Catalog Card No. 65–26889

Printed in Great Britain by Blackie & Son Ltd., Bishopbriggs, Glasgow.

(2294/66)

CONTENTS

GENERAL INTRODUCTION

THE volumes in this division have been planned to provide a comprehensive treatment of chemistry at the intermediate level—that is, the sixth-form/first-year university level. Readers are assumed to have a background of O-level chemistry and of O- or A-level physics and a working knowledge of elementary mathematics.

The books of the division will meet all the requirements of the recently revised A-level syllabuses of the examining boards and an attempt has been made to anticipate the nature of future revision of these syllabuses. They will also cover the ground for university scholarships and for first-year university examinations, such as those set to intermediate, medical, engineering students, etc. They will provide ordinary national certificate students in technical colleges with all they need and will constitute a useful background and companion to the studies of higher national candidates. In the U.S.A., first- and second-year college students will find them directly relevant to their studies and they will be of value to high-school students for reference purposes.

The present volume contains authoritative accounts of chemical kinetics and of surface and colloid chemistry—important topics which usually receive most unsatisfactory treatments in textbooks written at this level.

PREFACE

THIS book is about two subjects that do not yet bulk large in the school chemistry curriculum. The formal treatment presently called for in most examination syllabuses could be set out in far fewer pages. A good course should, however, leave the student aware that he is dealing with a living subject that is not fully understood. Both surface and colloid chemistry and chemical kinetics are among the most active fields of research. Accordingly, we have tried in our separate sections to outline some of the concepts that currently exercise workers in the field.

A. F. TROTMAN-DICKENSON

G. D. PARFITT

ACKNOWLEDGMENTS

I am grateful to Dr. R. C. Palmer and the Wool Industries Research Association, Leeds, for permission to reproduce Fig. 10. Also to my wife and Dr. A. L. Smith for a great deal of help in the preparation of the manuscript.

G. D. P.

FUNDAMENTAL CONSTANTS

Avogadro number N_0	$= 6{\cdot}023 \times 10^{23}$ mole^{-1}
Boltzmann constant $\mathbf{k}$	$= 1{\cdot}380 \times 10^{-16}$ erg deg^{-1}
Gas constant $\boldsymbol{R}$	$= 0{\cdot}082\,05$ l. atm deg^{-1} mole^{-1}
	$= 8{\cdot}314 \times 10^{7}$ erg deg^{-1} mole^{-1}
	$= 1{\cdot}987$ cal deg^{-1} mole^{-1}
Standard Gravitational Acceleration g	$= 980{\cdot}7$ cm sec^{-2}
Electronic charge e	$= 4{\cdot}803 \times 10^{-10}$ e.s.u.
$2{\cdot}303 \log_{10} x$	$= \log_e x$

BOOK I

Chemical Kinetics

BY A. F. TROTMAN-DICKENSON

CONTENTS

CHAPTER 1

THE KINETIC DESCRIPTION OF CHEMICAL CHANGE

THE heat of combustion of n-heptane is 1150 kcal mole^{-1}. The equilibrium position for the reaction

$$C_7H_{16} + 11O_2 = 7CO_2 + 8H_2O$$

therefore lies far over to the right. Petrol and its vapour can, however, remain in contact with air at room temperature for an almost infinite time provided that no spark ignites the mixture. Chemical kinetics is the science that provides the framework within which one can understand why the mixture remains unaltered, and how the rate of the reaction, when it occurs, depends upon the temperature and the concentration of oxygen. A complete kinetic description will detail the various reaction steps by which the ultimate products are formed and the rate of each step.

The importance of temperature and concentration in the promotion of chemical reaction has been recognized, if not explicitly stated, since chemistry was studied by the alchemists, but little advance could be made until the idea of a rate of a reaction had been suitably defined. Satisfactory descriptions of reaction rates were evolved less than one hundred years ago and the factors that determine these rates have only been recognized in the last forty or fifty years. It is the most fundamental task of the worker in chemical kinetics to determine these factors and to understand them in terms of the atomic nature of matter. This task is far from completed. The prediction of rates at which molecules react is much more hazardous than the prediction of their structure.

Only a small proportion of kinetic investigations are begun with the object of settling fundamental questions. Most work is done for other purposes, such as to determine the mechanisms of reactions. All the knowledge of reaction mechanisms which is treated under the heading of physical organic chemistry comes from studies of chemical kinetics in solution. Sometimes the results do not appear to be given in a conventional kinetic form, but it should be realized that the statement that the chlorination of n-butane yields 32 per cent n-butyl chloride and 68 per cent sec-butyl chloride at 300°C records the result of a kinetic experiment. Kinetic principles also determine the course of inorganic reactions, but this has not been so apparent because the rates of possible alternative reactions frequently differ by several orders of magnitude and because the rates of reactions between ions are frequently so great that only in the last few years have suitable methods of study been developed. In industry, kinetics is studied to determine the best sizes for reactors and the conditions that yield the greatest proportion of valuable products. The functioning of biological systems is also governed by relative reaction rates. Many of the most striking advances in our understanding of metabolism have come through the application of kinetic techniques to biochemistry.

It is convenient when discussing kinetics, to divide reaction systems into two classes: homogeneous and heterogeneous. Homogeneous reactions occur in one phase; heterogeneous reactions occur at the interface between two phases, one of which is usually a solid and the other a gas or a liquid. Heterogeneous systems will be considered in the last chapter.

Homogeneous reactions can themselves be classified under three headings:

(*a*) according to the number of molecules in the stoichiometric equation;

(*b*) according to the way in which their rates depend upon the concentration of the reactants;

(*c*) according to the number of molecules participating in the rate determining process.

The first classification is useful for the discussion of equilibrium constants but is not relevant to kinetics. The second classification is of great importance for the description of experimental results. The third is essential for the understanding of the factors that control the rates of reaction. It is convenient to consider now the second of these classifications.

The Rate of a Chemical Reaction. The rate of a chemical reaction is measured by analysing a reaction mixture at suitable intervals of time. It is tedious to repeat numerous conventional quantitative analyses, such as titrations, so that secondary indications of the composition of the mixture are frequently used. Thus the decomposition of gaseous t-butyl bromide

$$C_4H_9Br = C_4H_8 + HBr$$

can be followed by condensing the reaction mixture, adding water, and titrating the solution with standard alkali. It is more convenient to follow the change in pressure as it increases with the release of the extra molecule of hydrogen bromide. Although such secondary techniques are generally used, it is essential that mixtures are analysed so as to ensure that the pressure or other change results from the postulated chemical equation. Many early workers on gas phase reactions drew erroneous conclusions from results based on insufficient analyses. Not only must absolute methods of analysis be used but all results should be expressed in absolute terms, such as mole l^{-1} sec^{-1} (the number of moles of product formed or reactant removed from a litre of mixture per second). The rate of formation of a substance X is often written in the form $R(X)$ so for the decomposition of t-butyl bromide, the rate is given by

$$R(HBr) = R(C_4H_8) = -R(C_4H_9Br)$$

The rate can also be written in a differential form:

$$R(HBr) = d[HBr]/dt$$

The rate of a reaction is very rarely determined by direct experiment. As the differential form suggests it is obtained from the tangent drawn to a curve which is a plot of the variation of the composition of the reaction mixture with time. As can be seen from Fig. 1.1 the initial rate of reaction can be fairly accurately found from measurements where the composition of the reaction mixture has only been changed by a few per cent. Such a determination will only be accurate if the product and not the reactant is measured.

The Order of a Reaction. Experiments show that the rates of chemical reactions depend upon the concentrations of the reactants. For instance, a cigarette burns much more rapidly in pure oxygen at atmospheric pressure than in air. The rates of many reactions depend upon concentration in a simple way. Thus the rate of the reaction

$$\underbrace{\overline{CH_2CH_2CH_2}}_{\text{cyclopropane}} = \underbrace{CH_3CH{:}CH_2}_{\text{propylene}}$$

is given by $R(CH_3CH{:}CH_2) = k_1[\overline{CH_2CH_2CH_2}]$ [1]

where k_1 is the *rate constant* of the reaction for the particular temperature and the quantity in square brackets is a concentration. The isomerization of cyclopropane is described as a *first-order* reaction, because the rate of formation of product is proportional to the first power of the concentration of the cyclopropane. The rate is independent of the concentrations of all other substances that may be present, including the product.

An example of a *second-order* reaction is the combination of hydrogen and iodine:

$$H_2 + I_2 = 2HI \qquad [2]$$

This reaction was beautifully studied by M. Bodenstein (1896) who sealed up the reactants in glass bulbs at 0°C, and placed them in a thermostat, usually a vapour bath of sulphur, mercury or some other substance with a convenient boiling point. After a

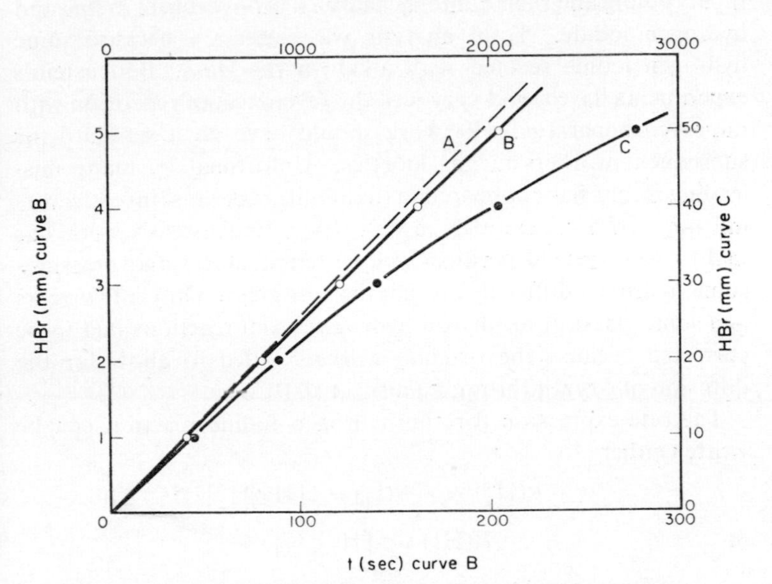

FIG. 1.1. The decomposition of t-butyl bromide. The rate of the reaction can be measured by determination of the hydrogen bromide released. Below are tabulated the times taken for the release of the stated pressure of hydrogen bromide from an initial 100 mm of t-butyl bromide at 250°C (the back reaction has been neglected). It can be seen from the figure that it is simple to determine accurately the initial rate of formation A of hydrogen bromide from the first table of results B but very difficult from the second table C. The reader can verify this statement by drawing large scale diagrams. Note that with the change in scales the initial slope should be the same for both curves B and C.

(B) HBr (mm)	t (sec)
0	0
1	40·2
2	80·7
3	122
4	164
5	206

(C) HBr (mm)	t (sec)
0	0
10	422
20	893
30	1417
40	2043
50	2776

known time the bulbs were removed, the reaction quenched by rapid cooling and their contents analysed for hydrogen, iodine and hydrogen iodide. Total analysis was necessary because some hydrogen iodide reacted with alkali in the glass. Bodenstein's experiments have stood very well the severe test of repetition with improved apparatus. His work should have set a standard for subsequent workers in gas kinetics. Unfortunately, many misleading results have appeared in the literature because investigators did not perform careful analyses. Even Bodenstein's work has had to be corrected because at high temperatures and pressures some hydrogen diffused through the soft glass. Only a few years ago some classical results on hydrogen atom reactions had to be corrected because the original workers failed to allow for the diffusion of oxygen through quartz at 800°C.

The rate expression for the hydrogen–iodine reaction can be written either

$$-R(H_2) = -R(I_2) = k[H_2][I_2]$$

or

$$R(HI) = k[H_2][I_2]$$

The second definition yields a rate constant that is numerically twice the first. The definitions are equally sound, but the difference shows that it is necessary to state clearly how a constant has been defined. The reaction is *second order* because its rate varies with the first power of the concentration of both reactants. The back reaction is also second order for its rate is given by

$$R(H_2) = R(I_2) = k_{-2}[HI]^2$$

or

$$-R[HI] = k_{-2}[HI]^2$$

and is proportional to the second power of the concentration of hydrogen iodide. The notation k_{-2} to denote the rate constant of the reverse of reaction [2] is particularly convenient in kinetics.

The dimensions of a rate constant vary with the order of its reaction. Thus for the isomerization of cyclopropane:

$$\text{rate (mole l}^{-1}\text{ sec}^{-1}) = k_1 \text{ [concentration] (mole l}^{-1})$$

and for the combination of hydrogen and iodine:

$$\text{rate (mole l}^{-1}\,\text{sec}^{-1}) = k_2\,[\text{concentration}]^2\,(\text{mole l}^{-1})^2$$

k_1, therefore, has dimensions sec^{-1} and k_2, $\text{mole}^{-1}\,\text{l. sec}^{-1}$.

Third-order reactions occur occasionally in the gas phase and more frequently in solution. An example is the reaction of nitric oxide with chlorine at low temperatures.

Then $$2NO + Cl_2 = 2NOCl \qquad [3]$$

$$-\tfrac{1}{2}R(NO) = -R(Cl_2) = \tfrac{1}{2}R(NOCl) = k_3[NO]^2[Cl_2]$$

The reader can readily verify that k_3 has the dimensions $\text{mole}^{-2}\,\text{l}^2\,\text{sec}^{-1}$. Some zero order reactions are known and some that accurately follow kinetics of fractional order, but the rates of many reactions that have well defined kinetics cannot be expressed in terms of order. For example, the rate of formation of hydrogen bromide in the combination of hydrogen and bromine is given by

$$R(HBr) = \frac{(k_a k_b/k_{-a})K^{\frac{1}{2}}[H_2][Br_2]^{\frac{1}{2}}}{k_b/k_{-a} + [HBr]/[Br_2]}$$

where k_a, k_b and k_{-a} are rate constants. The formation of hydrogen fluoride and hydrogen chloride from the elements does not follow any well-defined rate law, either simple or complicated. This variation of behaviour within the halogen family should serve as a warning to those who hope to draw analogies in kinetic behaviour.

The order of a chemical reaction should be regarded as a mathematical convenience, not as a fundamental property. The order can change if the reaction conditions are varied. The rate equation of the hydrolysis of ethyl acetate in aqueous solution

$$CH_3COOC_2H_5 + H_2O = CH_3COOH + C_2H_5OH \qquad [4]$$

can be written $$R(CH_3COOH) = k_4[CH_3COOC_2H_5]$$

The concentration of water is so large that the variation of its concentration cannot be detected. If the reaction were carried out in an inert solvent the equation would be

$$R(CH_3COOH) = k'_4[CH_3COOC_2H_5][H_2O]$$

and the constants would be related by the equation

$$k_4 = k'_4[H_2O]_{aq}$$

where $[H_2O]_{aq}$ is the concentration of water in a dilute aqueous solution, usually taken as 55·5 mole l^{-1}. This hydrolysis is sometimes referred to as a "pseudo-first-order" reaction. The terminology is confused and should be avoided. The only true criterion as to whether a reaction is of a particular order is the mathematical form in which the results are best cast. If first-order equations fit, then the reaction is first order.

The Algebra of Chemical Kinetics

The rate equation for a first-order reaction, A → products

$$-d[A]/dt = k[A] \text{ mole } l^{-1} \sec^{-1} \tag{1}$$

can be written

$$-d[A]/[A] = -d \ln [A] = k\, dt \tag{2}$$

If the initial concentration $[A]_0$ decreases to $[A]_t$ after t sec then

$$-\int_{[A]_0}^{[A]_t} d \ln [A] = k \int_0^t dt \tag{3}$$

$$\text{Since } \int d \ln [A] = \ln [A] + \text{constant of integration}$$

$$-\ln [A]_t + \ln [A]_0 = \ln ([A_0]/[A_t]) = kt \tag{4}$$

which can be arranged to the more convenient form

$$k = (1/t) \ln ([A]_0/[A]_t) = (1/t)\, 2{\cdot}303 \log ([A]_0/[A]_t) \tag{5}$$

Alternatively, we can write

$$[A]_t = [A]_0 \, e^{-kt} \tag{6}$$

The rate constant can be found by inserting into equation (5) the results of analyses for A (or equally well B) made at a series of times t. Alternatively, as can be clearly seen from equation (4), a plot of the logarithm of $[A]_0$ against t yields a straight line of slope $-k$. The first procedure is generally used when a run yields a single determination of the variation of concentration with time. This will occur when the entire contents of the reaction vessel are taken to provide sufficient material for analysis. The second procedure is used when a series of samples of one mixture are analysed or when a physical method of following the reaction yields a series of values. Since it is more convenient to use decadic rather than natural logarithms, the student must not overlook the necessity of multiplying the slope obtained by plotting decadic logarithms by the scaling factor of 2·303. There are many instances in the literature of authors leaving their results in decadic logarithms. No great harm is done if this is clearly stated, but the practice is best avoided. Older papers sometimes contain rate constants expressed in terms of minutes; the second is now to be preferred as a measure of time. The treatment of experimental results for a first-order reaction is shown in Example 1.1.

The simplest form of rate equation for a second-order reaction, say $A+B \rightarrow$ products, is found when the concentrations of A and B are the same; this usually occurs when the species are identical as in the decomposition of hydrogen iodide.

Then $\quad -d[A]/dt = k[A]^2 \quad$ or $\quad -d[A]/[A]^2 = k\,dt \qquad (7)$

since $$-\int \frac{d[A]}{[A]^2} = \frac{1}{[A]} + \text{constant of integration.}$$

$$\frac{1}{[A]_t} - \frac{1}{[A]_0} = \frac{[A]_0 - [A]_t}{[A]_0[A_t]} = kt$$

or $$k = \frac{1}{t}\frac{[A]_0 - [A]_t}{[A_0][A_t]} \text{ mole}^{-1}\text{ l. sec}^{-1} \tag{8}$$

The integration for the general case for which [A] ≠ [B] is more difficult, it yields

$$k = \frac{1}{t}\frac{1}{[A]_0-[B]_0}\ln\frac{[B]_0[A]_t}{[A]_0[B]_t} \tag{9}$$

It can be seen that if [A] and [B] are nearly but not exactly equal, the last term closely approaches unity. Reliable rate constants can then only be found from unnaturally accurate experiments. Normally rate constants are found by substitution in either equation (8) or (9). If equation (8) is rewritten in the form

$$kt = 1/[A]_t - 1/[A]_0 \tag{10}$$

it can be seen that a straight line of slope k can be obtained by plotting values of $1/[A]_t$ against the times. Equation (9) can be written

$$([A]_0-[B]_0)kt = \ln([A]_t/[B]_t) - \ln([A]_0/[B]_0) \tag{11}$$

and a line of slope $([A]_0-[B]_0)k$ obtained from a plot of the values of $\ln([A]_t/[B]_t)$ against t.

The special case for a third-order reaction in which

$$-d[A]/dt = k[A]^3$$

gives the result

$$k = \frac{1}{2t}\frac{1}{[A]_t^2}-\frac{1}{[A]_0^2}\ \text{mole}^{-2}\,\text{l.}\,\text{sec}^{-1} \tag{12}$$

The more general cases are complicated.

Not all reaction systems involve a single irreversible reaction that is rate determining. Systems such as the combination of hydrogen and iodine proceed to an equilibrium mixture containing a considerable fraction of the reactants. Others involve consecutive reactions such as are found in the series of products formed by the radioactive decay of uranium or thorium. Exact solutions have been found for the differential equations describing some of these phenomena; they are discussed in the larger texts on formal kinetics. Frequently, solutions can only be found by numerical analysis, preferably with the aid of a computer.

The order of a reaction is usually found by substituting experimental values of concentrations at different times into the integrated forms of the rate equations. If one of the equations yields k's that do not vary either with time when the reaction is carried to 75 per cent of completion or with changes in the initial concentration, then the correct order has been found.

Alternatively, the half-life method can be used. The half-life of a reaction is the time that must elapse before half the material initially present has been consumed. The half-life of a first-order reaction can be found by substituting $[A]_0/2$ for $[A]_t$ in equation (5) when

$$k = \frac{1}{t}\ln\frac{[A]_0}{([A]_0/2)} \quad \text{or} \quad t = \ln 2/k \text{ sec}$$

The half-life, which is often written $t_{\frac{1}{2}}$ is independent of the initial concentration. If we similarly substitute for $[A]_t$ in equation (8) for a second-order reaction

$$t_{\frac{1}{2}} = 1/k[A]_0 \text{ sec}$$

and the half-life is inversely proportional to the initial concentration. No simple meaning can be given to the half-life of a second-order reaction of substances at different concentrations as described by equation (9). For a third-order reaction following equation (12), we have $t_{\frac{1}{2}} = 3/(2k[A]_0^2)$ sec. The half-life of a reaction is found by interpolation from a plot of reactant concentration against time which should be run until the reaction is 80 per cent complete. Inspection of determinations at several initial concentrations will show which expression best fits the results. The three-quarter-life can then also be derived; it is, of course, twice the half-life for a first-order reaction, but not for reactions of other orders.

Other ingenious methods of determining order are described in the larger texts but they are rarely used except as exercises for the student.

Molecularity

A chemical reaction may occur as simply as its stoichiometric equation suggests, in which case it is described as elementary. Frequently, reactions occur in a series of steps and can involve species that are not represented in the equations. By careful tests, many of which will be kinetic, it should be possible to write mechanisms for the overall reactions in terms of steps each one of which is elementary. The definition of an elementary reaction is therefore very similar to the old definition of an element: an elementary reaction is a reaction that cannot be resolved into simpler reactions by chemical means. It can, of course, be described more fundamentally in terms of electrons and nuclei. The proof that a reaction is elementary is also similar to the proof that a substance was an element before the concept of atomic number was developed. It relies upon the application of some well established tests and the absence of conflicting evidence.

The term molecularity can only be used to describe elementary reactions. A unimolecular reaction is one which, ignoring problems of transfer of energy, could occur if only one molecule were present in an isolated system. A bimolecular reaction requires two molecules, which may be either like or unlike. A termolecular reaction requires three. A bimolecular reaction is necessarily second order, though the converse is not true. Order merely describes the mathematical form of the results. The difference between order and molecularity must be stressed because both concepts are useful when properly applied; unfortunately, they have frequently been confused in the past.

Examples of unimolecular reactions include the decomposition of t-butyl bromide and the isomerization of cyclopropane described above. Also unimolecular are the decomposition of cyclobutane to ethylene and of acetic anhydride to acetic acid and ketene. Simple bimolecular reactions include the hydrogen–iodine reaction (but not, as could be deduced from the remarks on their order, the reaction of hydrogen with bromine, chlorine or fluorine), and the hydrogen iodide decomposition over a very wide range of temperature. Other reactions are

$$NO_2 + CO = NO + CO_2$$
$$NO_2Cl + NO = NO_2 + NOCl$$
$$NO + O_3 = NO_2 + O_2$$

Examples of unimolecular reactions in solution are quoted in the literature. If ions are not involved the attribution can be accepted, for several unimolecular gas phase reactions proceed at the same rate in a solvent. Ions, however, are heavily solvated and this solvation must change as the reaction proceeds. This concerted movement should not be classified as unimolecularity. An appropriate test is to see whether a change in solvent effects the rates, if it does, the molecularity of the reaction is doubtful. Termolecular reactions may not exist in the gas phase. The classic examples are all reactions of nitric oxide:

$$2NO + O_2 = 2NO_2$$
$$2NO + Cl_2 = 2NOCl$$
$$2NO + Br_2 = 2NOBr$$

Nitric oxide forms a dimer and it may be this dimer which reacts. Similarly the combination of atoms such as

$$I + I + Ar = I_2 + Ar$$

probably involves the prior association of two of the atoms and cannot be said for certain to occur in one step.

EXAMPLE 1.1. *The calculation of first-order rate constants*

The table below gives the amount of 1-methyl cyclobutene remaining in a reaction vessel at 179·4°C after various times. The reaction is

$$\begin{array}{l} CH_2-C-CH_3 \\ | \quad\;\; \| \\ CH_2-CH \end{array} = CH_2{=}CH-\overset{\displaystyle CH_3}{\overset{|}{C}}{=}CH_2$$

t (sec)	180	300	420	600	900	1200	1500
% methylcyclo-butene remaining	86·9	79·5	72·2	62·3	49·4	38·6	30·6

If these results are inserted into the equation

$$k = (1/t)\log(2{\cdot}303 \times 100/\%\ \text{remaining})$$

we get

10^4k (sec^{-1})	7·80	7·84	7·76	7·89	7·84	7·93	7·90

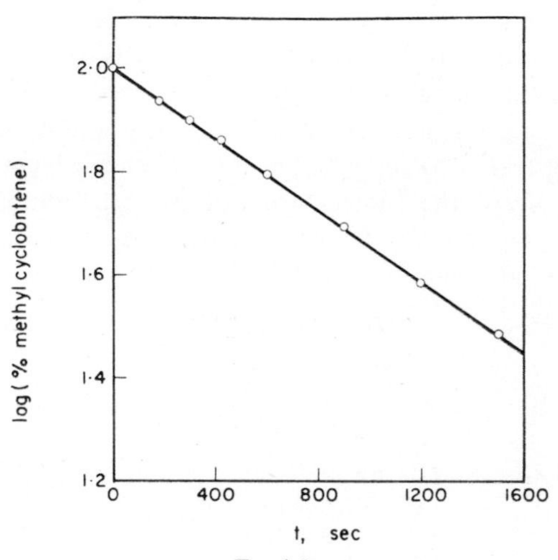

Ex. 1.1.

from which we find the average rate constant $7{\cdot}85 \times 10^{-4}$ (sec^{-1}). Alternatively the rate constant can be found graphically. The logarithm of the percentage remaining should then be tabulated as below.

log (%)	1·939	1·900	1·859	1·795	1·694	1·587	1·486

The slope of the graph is given by

$$(2{\cdot}000 - 1{\cdot}450)/1600 = 3{\cdot}438 \times 10^{-4}$$

Hence

$$k = 2{\cdot}303 \times 3{\cdot}438 \times 10^{-4}$$
$$= 7{\cdot}92 \times 10^{-4}\ \text{sec}^{-1}$$

The difference between the two values comes from slight graphical errors. Note, it is usually most convenient to plot decadic logarithms and convert to natural logarithms in the last step.

EXAMPLE 1.2. *The dimerization of trifluorochloroethylene—a second-order reaction*

The dimerization of trifluorochloroethylene:

$$\begin{matrix} CF_2{=}CFCl \\ + \\ CF_2{=}CFCl \end{matrix} = \begin{matrix} CF_2-CFCl \\ | \qquad | \\ CF_2-CFCl \end{matrix}$$

is one of the very few reactions of the type $2A = B$ that can be fully described by the equation $R(B) = k[A]^2$. Its kinetics have been simply studied by measuring the change in pressure in a

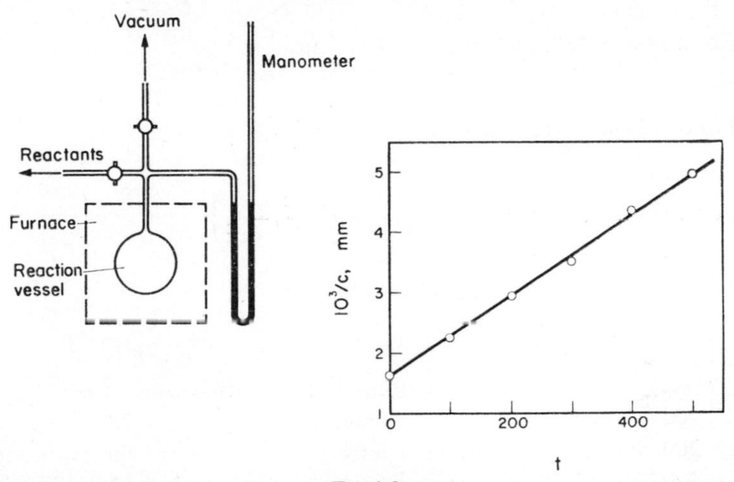

Ex. 1.2.

reaction vessel with time. The apparatus shown in the figure (Ex. 1.2) is suitable for this purpose. The pressure readings (p mm) obtained at 440·4°C and an initial pressure of 620 mm are those given in the second column of the table below. These must be converted into partial pressures of the reactant (c mm) by use of the equation:

$$c = 2p - c_0$$

where c_0 is the initial pressure of reactant, here 620 mm

t (sec)	p (mm)	c (mm)
0	620	620
100	533	445
200	480	340
300	453	285
400	425	230
500	411	202

The rate constant can be calculated for each value of t. The appropriate equation is that given below.

For a first-order reaction: $$k = \frac{1}{t}\ln\frac{c_0}{c}$$

For a second-order reaction: $$k = \frac{1}{t}\left(\frac{1}{c} - \frac{1}{c_0}\right)$$

For a third-order reaction: $$k = \frac{1}{2t}\left(\frac{1}{c^2} - \frac{1}{c_0^2}\right)$$

Hence we obtain the values listed below:

	First	Second	Third
t (sec)	$10^3 k$ (sec^{-1})	$10^6 k$ (mm^{-1} sec^{-1})	$10^8 k$ (mm^{-2} sec^{-1})
100	3·32	6·34	1·28
200	3·00	6·64	1·50
300	2·59	6·63	1·62
400	2·48	6·84	2·04
500	2·24	6·61	2·19

It can be seen that the "constant" calculated from the assumption of second-order kinetics is much the most constant. This is good proof that the reaction is second order. It should be noted that the "constant" of lower order falls with time whereas that of higher order rises. This can be a useful indication when results are to be fitted to an equation by trial and error.

Once the order of a reaction is known, rate constants are best found graphically as shown in the figure, where $1/c$ is plotted against t. The rate constant is then equal to the slope.

$$\begin{aligned} k &= 10^{-3}(4{\cdot}950 - 1{\cdot}613)/500 \\ &= 6{\cdot}66 \times 10^{-6}\,\text{mm}^{-1}\,\text{sec}^{-1} \end{aligned}$$

CHAPTER 2

THE EFFECT OF TEMPERATURE AND LIGHT

The Effect of Temperature

The rates of most organic reactions are strikingly increased by an increase in temperature; commonly a rise of 10°C doubles the rate constant. The dependence of many rate constants on temperature can be expressed in terms of the Arrhenius equation:

$$k = A\,e^{-E/\boldsymbol{R}T} \quad \text{or} \quad \ln k = \ln A - E/\boldsymbol{R}T$$

which can also be written

$$k = A\exp(-E/\boldsymbol{R}T) \quad \text{or} \quad \log k = \log A - (E/2{\cdot}303\boldsymbol{R}T)$$

In these expressions $\boldsymbol{R}$ is the gas constant, which should be in units of cal mole^{-1} deg^{-1}. (It has the numerical value 1·987 which for most purposes can be taken as 2.) T is the absolute temperature while A and E are the Arrhenius parameters; A is best called the A factor and E the Arrhenius activation energy. The name activation energy is intended to imply that it is the energy which the reacting molecules must acquire before they can react. In this book the term will be used without the qualifying "Arrhenius". Readers of other works should be careful to note whether the writers have used the Arrhenius definition. Frequently other definitions are used which give numerical values that differ by several thousand cal mole^{-1}; occasionally the writer does not state which definition he has adopted.

The activation energy of a reaction can be calculated from the Arrhenius equation if the rate constant has been determined at two different temperatures (Ex. 2.1). It is better to measure the rate constant at several temperatures and to obtain E by graphical methods, as is shown in Example 2.2. When the Arrhenius equation is written in the form

$$\log k = \log A - (E/2{\cdot}303\boldsymbol{R}) \,.\, 1/T$$

it can be clearly seen that a plot of $\log k$ against the reciprocal of the absolute temperature is a straight line. The intercept is $\log A$ and the slope $-E/2{\cdot}303\boldsymbol{R}$ or $-E/4{\cdot}574$. Frequently it is inconvenient and inaccurate to extrapolate to $1/T = 0$ to find the intercept, and $\log A$ is found by substitution in the equation.

If the variation of the rate of a reaction with temperature does not approximately fit the Arrhenius equation, then it is certain that the reaction is not elementary. The converse is not true. The rate constants of many complex reactions fit the equation well.

The Effect of Light

Provided that the temperature is constant, the rates of chemical reactions are affected little by the environment. Pressures of hundreds of atmospheres only alter rate constants by a few per cent, though of course the rate of a gas reaction is raised by the increase in concentration. Nor are rates affected by large electric or magnetic fields. There are, however, many chemical systems that react either faster or by alternative mechanisms under the influence of light. If light is to affect a reaction mixture it must be of such a wavelength that it is absorbed by one of the constituents. This principle will appear obvious after a moment's thought, but it is easily forgotten. When originally enunciated before chemical matters were well understood, the principle was accorded the dignity of a law named after its promulgators, Grotthus (1817) and Draper (1841). Few small molecules absorb visible light (between 4000 and 8000 Å wavelength). Almost all absorb infrared radiation (greater than 8000 Å) but this has no photochemical effects. Ultraviolet radiation with a wavelength between 4000 and 2000 Å is most often used.

The chemical consequences of the absorption of light can be

dramatic. An example is the explosion of mixtures of hydrogen and chlorine which do not react in the dark at room temperature. They are also extremely important, for all higher forms of life depend upon photosynthesis. Einstein's law of photochemical equivalence (1905) which provides the theoretical basis for the systematic treatment of the subject, states that any molecule or atom activated by light absorbs only one quantum of light, and this causes activation. The energy absorbed by one molecule with one quantum of light of frequency, ν, is $h\nu$ where h is Planck's constant ($6{\cdot}63 \times 10^{-27}$ erg sec). The energy absorbed by one mole is then $Nh\nu$, where N is the Avogadro's number ($6{\cdot}023 \times 10^{23}$); thus

$$E = Nh\nu = Nhc/\lambda$$

where c is the velocity of light ($2{\cdot}998 \times 10^{10}$ cm sec^{-1}) and λ is the wavelength of the light in centimetres. In more familiar units with λ in Ångstroms

$$E = \frac{2{\cdot}85 \times 10^{5}}{\lambda} \text{ kcal mole}^{-1}$$

The energy of a quantum of visible light of 4000 Å is thus 71 kcal mole^{-1} which is just sufficient to break a weak covalent bond. The principal line from a low-pressure mercury arc (which is frequently used) is 2537 Å or 112 kcal mole^{-1}. This is sufficient to break all but a few single covalent bonds. A quantum of X-rays with a wavelength of about 1 Å contains energy which, if it could be used efficiently, would disrupt 3000 molecules. Nuclear radiation is even more energetic, but its chemical effects are similar to those of X-rays, and the two are considered together as "Radiation Chemistry". Radiation chemistry has aspects of kinetic interest but they lie outside the range of this book.

We shall be concerned with only one of the several consequences of the absorption of ultraviolet and visible light: the breaking of covalent bonds. When methyl iodide is photolysed in a system such as that shown in Fig. 2.1 it is split into a free methyl radical and an iodine atom

$$CH_3I + h\nu = CH_3 + I$$

Free radicals and atoms are extremely reactive; they combine together and abstract atoms from stable molecules. Simple photolytic systems are therefore rare, although the primary photolytic acts of dissociation are straightforward.

An important class of photolytic phenomena is the initiation of a sequence or chain of reactions, such as occurs in the hydrogen–chlorine explosion. The absorption of one quantum of light results in the formation of many molecules of hydrogen chloride. The

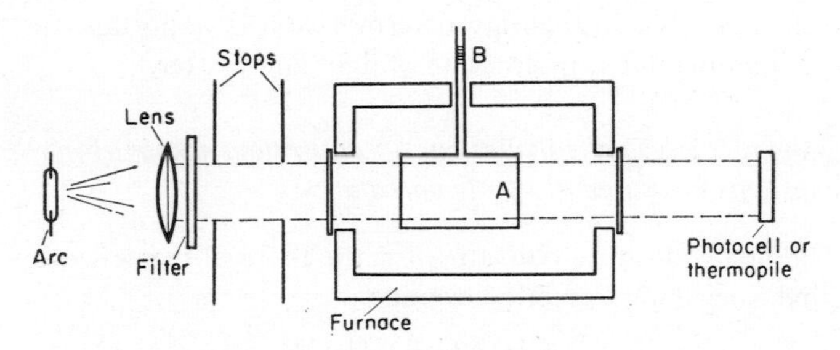

FIG. 2.1. Basic apparatus for investigation of the photolysis of compounds such as methyl iodide, acetone, etc. The compound to be photolysed is contained in the quartz cylindrical reaction vessel A which is usually about 10 cm long and 5 cm diameter with optically flat windows sealed to either end. It is connected to a gas handling apparatus through the side arm by way of the graded quartz–borosilicate glass seal B. The reaction vessel is placed in a furnace with plane quartz windows at either end.

The light source is normally a medium pressure mercury arc of 125 to 500 W. A convenient source is the normal commercial arc sold for lighting purposes with the outer glass envelope removed. Such arcs give fairly intense ultraviolet light at wavelengths down to about 2400 Å. Ultraviolet light quickly damages eyes so the arc must be carefully shielded.

The light is collimated by a quartz lens and the two stops so that the beam is reasonably parallel. A filter can be inserted if it is intended to work with a limited range of wavelengths. Screens can be inserted into the optical system to reduce the intensity. The light passing out of the cell falls on a photocell or photomultiplier, which can be calibrated to indicate the number of quanta absorbed in the cell. Direct measurement of the energy passing through the cell by means of a thermopile is rarely attempted unless absolute measurements are indispensable.

quantum yield is then said to be large. It is defined for a system by the relation

$$\text{Quantum yield} = \frac{\text{Number of molecules transformed}}{\text{Number of quanta of actinic light absorbed}}$$

as illustrated in Example 2.3. Einstein's law can therefore be restated by saying that the quantum yield of a single elementary process is unity. Numerical values of quantum yields should only be stated for monochromatic light of a specified wavelength, though frequently the values change little over a fair range of wavelengths. Their accurate determination is important for the understanding of the interaction of light and matter.

EXAMPLE 2.1. *The calculation of activation energy from rate constants measured at two temperatures*

The first-order rate constants for the thermal isomerization of methyl isocyanide to methyl cyanide

$$CH_3NC = CH_3CN$$

have been found to be $75{\cdot}0 \times 10^{-6}\ \text{sec}^{-1}$ at 199·4°C and $90{\cdot}8 \times 10^{-5}$ sec^{-1} at 230·4°C when the total pressure in the reaction vessel is 8 atmospheres. Use the relation

$$\log k = \log A - E/2{\cdot}303\boldsymbol{R}T$$

$$\log k_1 - \log k_2 = \frac{E}{2{\cdot}303\boldsymbol{R}}\left(\frac{1}{T_2} - \frac{1}{T_1}\right)$$

T°K	$10^3/T$	$\log k$
199·4+273·1	2·1164	−4·1249
230·4+273·1	1·9861	−3·0419

$$E = \frac{2{\cdot}303\boldsymbol{R} \times 10^3 \times (4{\cdot}1249 - 3{\cdot}0419)}{(2{\cdot}1164 - 1{\cdot}9861)}\ \text{cal mole}^{-1}$$

$$= 38{\cdot}02\ \text{kcal mole}^{-1}$$

A can simply be found by insertion in the equation.

Note that reciprocal tables should be used to find $10^3/T$.

In general, the greater the separation of the two temperatures the more accurate will be the result.

EXAMPLE 2.2. *The determination of activation energy by graphical means*

The results given in the first two columns of the table below have been reported for the rate of isomerization of 1,2-dimethylcyclobutene to 2,3-dimethylbuta-1,3-diene at various temperatures and a pressure of 3·0 mm.

$$\begin{array}{l} CH_2\text{—}C\text{—}CH_3 \\ \;| \qquad \| \\ CH_2\text{—}C\text{—}CH_3 \end{array} \qquad \begin{array}{c} CH_3 \qquad\qquad CH_3 \\ \diagdown \qquad \diagup \\ CH_2{=}C\text{—}C{=}CH_2 \end{array}$$

T	10^4k (sec^{-1})	$\log k$	$10^3/T$
149·7	0·162	$\bar{5}$·2095	2·366
159·3	0·435	$\bar{5}$·6385	2·313
163·1	0·607	$\bar{5}$·7832	2·293
168·3	0·991	$\bar{5}$·9961	2·266
170·7	1·246	$\bar{4}$·0948	2·253
173·3	1·58	$\bar{4}$·1987	2·240
179·0	2·59	$\bar{4}$·4133	2·212
182·2	3·50	$\bar{4}$·5441	2·197
187·5	5·58	$\bar{4}$·7466	2·173
193·3	8·91	$\bar{4}$·9499	2·144
197·0	12·25	$\bar{3}$·0881	2·128

Note the convenience of keeping $\log k$ in the above form. From the graph on p. 26,

$$\text{slope} = -(\bar{3}\cdot100-\bar{5}\cdot140)/2\cdot375-2\cdot125)$$
$$= -7\cdot840$$

$$\text{activation energy} = 2\cdot303R\times7\cdot840$$
$$= 35\cdot85 \text{ kcal mole}^{-1}$$

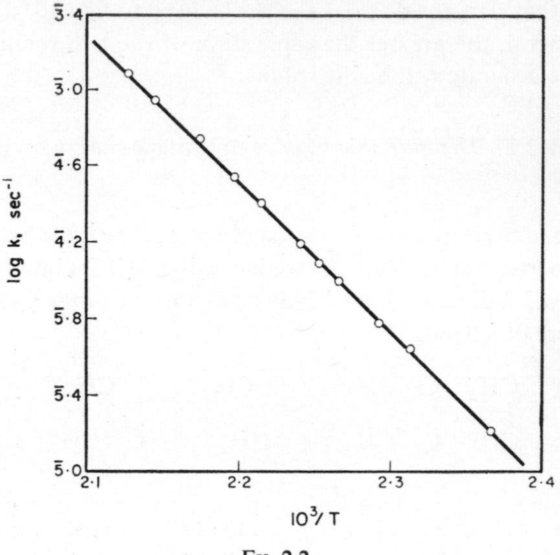

Ex. 2.2.

EXAMPLE 2.3. *The calculation of quantum yields*

When hydrogen iodide is photolysed with light of 2537 Å, it is decomposed to hydrogen and iodine

$$2HI = H_2 + I_2$$

In an experiment it was found that the absorption of $3{\cdot}07 \times 10^9$ ergs of energy measured on a thermopile decomposed $1{\cdot}30 \times 10^{-3}$ moles of hydrogen iodide determined by measurement of the hydrogen released.

The energy of a quantum of ultraviolet light of 2537 Å is

$$\begin{aligned} h\nu &= hc/\lambda \\ &= (6{\cdot}62 \times 10^{-27})(3{\cdot}0 \times 10^{10}/2{\cdot}537 \times 10^{-5}) \\ &= 7{\cdot}84 \times 10^{-12} \text{ ergs} \end{aligned}$$

The number of quanta absorbed by the hydrogen iodide is therefore:

$$3{\cdot}07 \times 10^{9}/7{\cdot}84 \times 10^{-12} = 3{\cdot}92 \times 10^{20}$$

If one molecule of hydrogen iodide is decomposed for each quantum absorbed, the amount of decomposition should be

$$3{\cdot}92 \times 10^{20}/6{\cdot}02 \times 10^{23} = 6{\cdot}52 \times 10^{-4} \text{ moles}$$

The quantum yield for the photolysis of hydrogen iodide is therefore

$$(1{\cdot}30 \times 10^{-3})/(6{\cdot}52 \times 10^{-4}) = 1{\cdot}99$$

CHAPTER 3

ELEMENTARY KINETIC SYSTEMS

Collision Theory of Bimolecular Reactions: Collisions

It has been established that all observed reactions are the result of one or more elementary reactions. Therefore, if the factors that control the rates of elementary reactions are understood, the behaviour of all systems can in principle be predicted. We shall now consider what can be said about these factors if we take the universal applicability of the Arrhenius equation and a knowledge of the parameters, as the only information derived from kinetic studies.

By a brief calculation based only on a knowledge of the approximate density of crystals and of Avogadro's number (the number of molecules in a gramme–molecule of material which is $6{\cdot}023 \times 10^{23}$), the reader can show that the distances that separate atoms in molecules must be 1 to 3 Å. The fact that the change in volume that accompanies melting is small shows that about the same distance separates molecules in all condensed phases. From this we can deduce that chemical bonding forces exert their maximum effect between nuclei separated by 1 to 3 Å. This effect must fall off very rapidly as the distance is increased, or all matter would be one big molecule. Much more detailed information can be obtained, particularly from X-ray crystallography and infrared spectroscopy, but the simple calculation is good enough for our purpose. It shows that before a gaseous bimolecular reaction can occur two molecules must approach each other so closely that they can be said to collide. Only then can an oxygen atom in nitrogen dioxide be attracted sufficiently strongly by the carbon atom in carbon monoxide to be broken away from the nitrogen, as it is in the reaction

$$CO + NO_2 = CO_2 + NO$$

Long before chemical kinetics was recognized as a discipline, the mathematical description of molecular collisions had been worked out by Maxwell and Boltzmann. It is known as the kinetic theory of gases. Starting from the premise that gases are composed of hard spherical molecules moving chaotically, it derives the rate of collisions between molecules of, say, nitrogen dioxide and carbon monoxide, as:

$$\begin{aligned}\text{Rate of collisions}\\ &= Z[NO_2][CO]\\ &= \tfrac{1}{4}(\sigma^2_{NO_2} + \sigma^2_{CO})\{8\pi RT(M_{NO_2} + M_{CO})/M_{NO_2}M_{CO}\}^{\frac{1}{2}}\\ &\qquad \times [NO_2][CO]\,.\,10^{-3}\,\text{mole}\,l^{-1}\,\text{sec}^{-1}\end{aligned}$$

where σ_{NO_2} is the collision diameter of nitrogen peroxide and Z is called the collision constant, M_{NO_2} is the mass of a molecule; the factor of 10^{-3} is inserted simply to convert to litres, for molecular dimensions are commonly measured in centimetres. Kinetic theory collision diameters only increase with the cube root of the number of atoms in the molecules so that for molecules of moderate complexity they vary little. The term involving the masses of the molecules is a square root and so is also fairly constant. Hence the rate of collision has a roughly constant value of $10^{11\cdot4}$ mole l^{-1} sec^{-1} and Z is $10^{11\cdot4}$ mole^{-1} l. sec^{-1}. The connexion of Z with a second-order rate constant can be seen from the similar dimensions of the two quantities. It is the rate constant that all bimolecular reactions would have if energetic and spatial considerations did not have to be satisfied. Occasionally these considerations are unimportant, and rate constants of this size are found for a very few reactions such as that of a sodium atom with a chlorine molecule

$$Na + Cl_2 = NaCl + Cl$$

and for the simplest of all collision processes, the transfer of translational energy and of rotational energy between large molecules.

Energy. The collision constant varies only with the square root of the absolute temperature. Experiment shows that the rate constants of most reactions have a large exponential dependence upon temperature. It is clear, therefore, that contact between molecules is not a sufficient although it must be a necessary condition for reaction. The exponential dependence occurs throughout physical chemistry, as in the variation of the vapour pressure of a liquid with temperature. The heat of vaporization can be regarded as the energy required to overcome the attractive forces that bind molecules of a liquid together. Similarly activation energy is the energy that the colliding molecules must possess before reaction can occur. If it is granted that this energy is needed, two questions present themselves. First, why does the energy involved appear with the temperature in an exponential relationship? Second, what determines the magnitude of the energy requirement?

The first question can only be fully answered in terms of a rather complicated argument, but much understanding can be gleaned from the consideration of some simple ideas. Let us consider the distribution of energy among atoms in a monatomic gas. This example is chosen because an atom can only possess translational energy. According to the simplest form of the kinetic theory of gases, all the atoms are assumed to have the same velocity. In fact it can be shown by the time-of-flight experiment described in Example 3.1 (see pp. 47–8), that the atoms have a wide range of velocities.

EXAMPLE 3.1

Zartman (1931) demonstrated the spread of atomic velocities with the apparatus shown in the figure. Bismuth atoms emerge from the furnace at 850°C through slit A. The beam is collimated by slit B. Slit C is cut in the side of a drum of 10 cm diameter which rotates at 120 or 240 rev/sec. When slit C is momentarily in line with slits A and B, a group of atoms enters the drum. An atom that travelled infinitely fast would hit a point on the drum diametrically opposite C. During the time that a slower moving atom crosses the drum, the drum will have moved forward so that the atom condenses on the glass slide. At the end of the run the slide is removed and the variation in density of the deposit found with a photometer. It is a simple matter to calculate, from the geometry of the system, what velocity an atom must have to hit a given point on the slide.

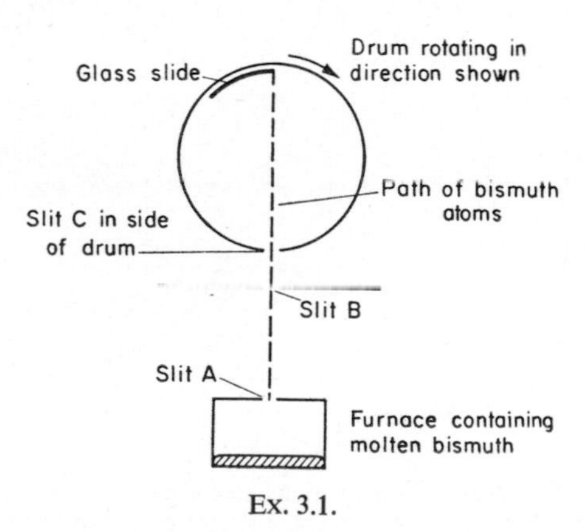

Ex. 3.1.

The result of this experiment can be represented best by Fig. 3.1. An experiment with the furnace at one temperature yields the curve T_1, at a higher temperature curve T_2 is obtained. For the present purpose the most important property of the distribution curves is the variation with temperature of the proportion of molecules whose velocities exceed a certain high limit. When this velocity is several times the mean velocity for T_1, the proportion of molecules above the limit increases very rapidly with temperature. It can be shown that the increase is exponential.

For a bimolecular reaction to occur the two molecules must be in contact and must together satisfy an energy requirement. The exponential form found in the Arrhenius equation and in the dependence of the number of atoms with velocities above a certain limit suggests a connexion. The principle of equipartition of energy requires that the dependence of the number of molecules

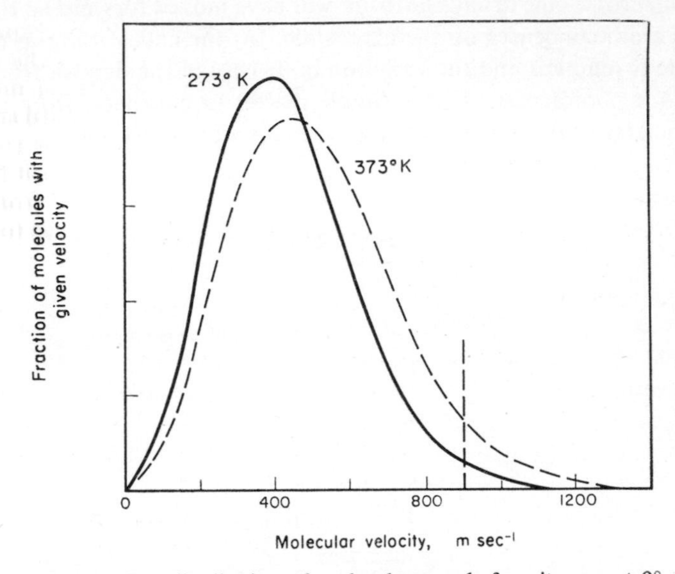

FIG. 3.1. The distribution of molecular speeds for nitrogen at 0° and 100°C. The molecular kinetic energies are proportional to the squares of the velocities. The important point to notice is that while the increase in temperature produces a comparatively small shift in the average velocity of the molecules the proportion of molecules with velocities greater than 900 m sec^{-1} has increased many times. The relation of this behaviour to the rates of chemical reactions can be seen from the fact that, in many reactions occurring around room temperature, the rate is doubled for a 10° rise in temperature. In other words 10° doubles the number of collisions that occur with more than the critical energy. On the other hand, around 300°K, 10° only increases the average energy of all the molecules by about 3 per cent.

with energy above the defined limit is the same for other degrees of freedom as it is for the translational one that we have considered. But we need not concern ourselves with this. We have adopted a collisional model. It is most reasonable to suppose that reaction will only occur if the collision is sufficiently violent. The energy dissipated is the measure of this violence. The mutual kinetic energy along the line of centres of the two molecules must be calculated. This is translational energy along an axis defined by the centre of the two molecules. Calculation shows that the fraction of collisions in which the energy exceeds a value E is given by $\mathrm{e}^{-E/RT}$. Hence the rate of collisions between nitrogen dioxide and carbon monoxide in which energy of collision exceeds E is

$$Z[\mathrm{NO_2}][\mathrm{CO}]\,\mathrm{e}^{-E/\;T}$$

which is the form of the Arrhenius equation. The fraction of collisions in which the energy is to be found not only in mutual kinetic energy but also in the vibrations of the molecules can also be calculated. The expression found is similar but not identical in that it contains terms that involve a non-exponential dependence on temperature. Experiments are rarely accurate enough to distinguish between the two models. In such a case one relies on the pragmatic principle of preferring the simplest hypothesis that will fit to within experimental error. Research workers are constantly on the look out for well defined reactions which would enable us to decide the relative merits of various models.

Unimolecular Reactions

The universality of the Arrhenius equation suggests that a similar molecular model should serve for all kinds of reactions, but unimolecular reactions present a difficulty. A single activated molecule has a finite probability of reaction, but it can only acquire its energy of activation by collision. Collisions are second-order processes. How therefore can the first-order rate constants be explained? Lindemann and Christiansen independently proposed a solution in 1921. They suggested that activated molecules were formed in energetic collisions (1) as is shown in Fig. 3.2. The activated molecule will, if left undisturbed for a given time (usually referred to as its lifetime), react (2) to give products. But at high pressures most of the activated molecules will be deactivated by collision (3).

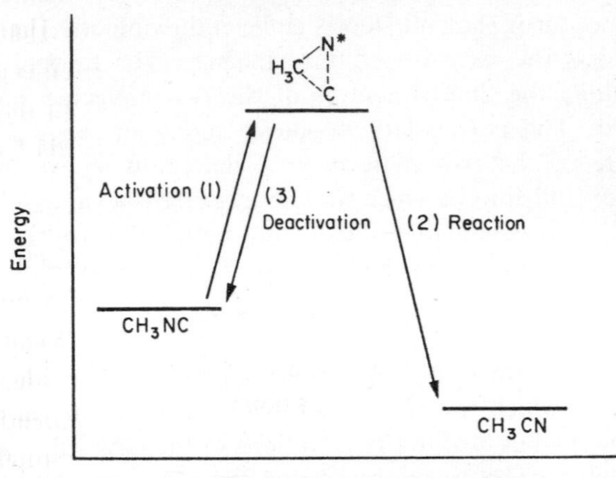

FIG. 3.2. Diagram representing the isomerization of methyl isocyanide according to the Lindemann scheme. The processes are as set out below:

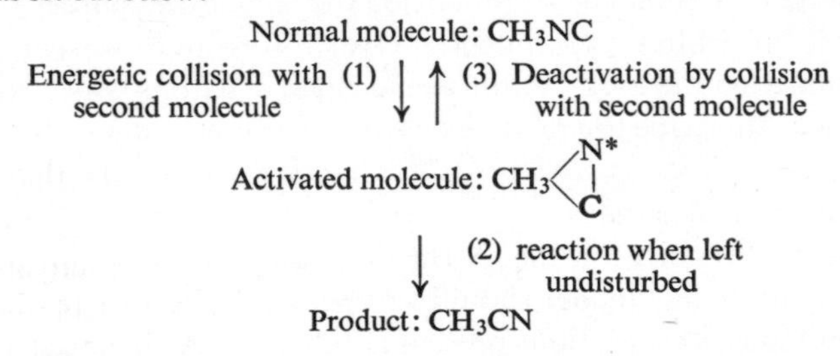

At high pressures the rates of reactions (1) and (3) will be much faster than the rate of reaction (2). Equilibrium between (1) and (3) will be established and a constant fraction of the molecules will be activated. This fraction will be large if the temperature is high. For a given compound it is proportional to $e^{-E/RT}$. When the lifetime is long compared with the interval between collisions

$$\begin{aligned}\text{Rate of formation of products} &= [\text{Activated molecules}]k_2\\ &= [\text{Total molecules}]k_2\,\kappa\,e^{-E/RT}\end{aligned}$$

where κ is a proportionality constant.

The first-order rate constant is defined by

$$\text{Rate of formation of products} = [\text{Total molecules}]k$$

$$\therefore\ k = k_2 \kappa \mathrm{e}^{-E/RT}$$

At low pressures, the chance that an activated molecule is deactivated before it forms products is small; the interval between collisions is long. When every molecule that is activated subsequently reacts, we have

$$\begin{aligned}\text{Rate of formation of products} &= \text{Rate of activation}\\ &= [\text{Total molecules}]^2 k_1\end{aligned}$$

This expression implies that second-order kinetics should be observed. In fact, the results obtained in each experiment fit first-order equations. This occurs because a reactant molecule can be activated either by collision with a reactant or a product molecule (or equally by molecules of any inert gas that is present). The total number of molecules in the system usually alters by less than a factor of two during a run. It can be taken as constant. At low pressures, therefore, we observe first-order kinetics within

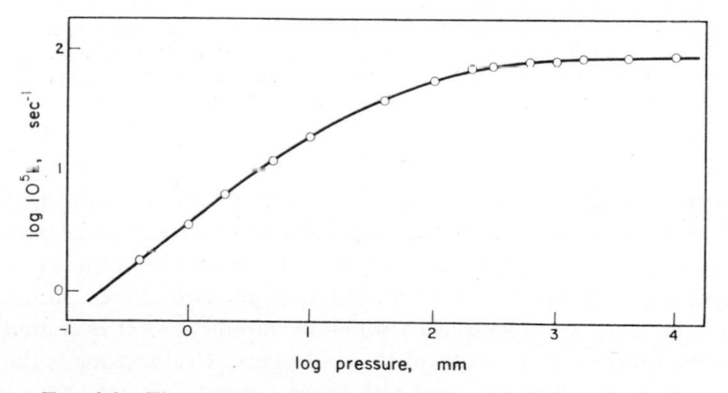

FIG. 3.3. The variation of the first-order rate constant for the isomerization of methyl isocyanide with pressure at 230·4°C. It can be seen that at the highest pressures the rate constant is independent of pressure. At the lowest pressures, k is proportional to p (the slope of the log–log plot is unity).

each run but the first-order rate constants are proportional to the initial pressure in the reaction vessel. The change in the rate constants for individual runs is best plotted against the initial pressure for the runs. Figure 3.3 shows this for the isomerization of methyl isocyanide to methyl cyanide. Figure 3.4 also shows the

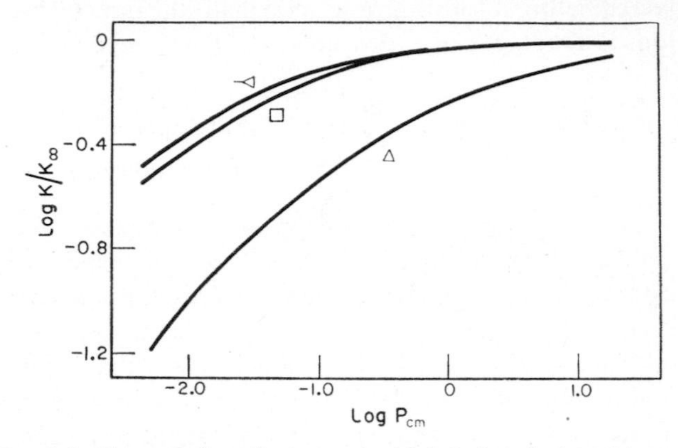

FIG. 3.4. The variation of rate constant (plotted as the logarithm of k as a fraction of the rate constant at infinite pressure) with pressure for (*a*) the isomerization of cyclopropane to propylene, (*b*) the decomposition of cyclobutane to two molecules of ethylene, and (*c*) the isomerization of methyl cyclopropane to various butenes. Note the similarity of the pressure dependence for the isomers methyl cyclopropane and cyclobutane. The Arrhenius parameters are similar for all three reactions.

variation of the rate constant for the isomerization of cyclopropane to propylene and for the decomposition of cyclobutane with pressure. The Arrhenius parameters for the reactions of these two molecules are similar. The separation of the two curves on the pressure axis is not caused by different parameters. It is caused by the different complexity of the molecules. Cyclobutane is the more complex. The activated cyclobutane molecule has the longer lifetime (the lower reaction rate) because in a complex molecule it takes longer before the critical arrangement for reaction is achieved. Methylcyclopropane can similarly be seen to have its normal high pressure rate constant down to a lower pressure than cyclopropane. The lifetimes of activated molecules are still the subject of active research. It is one of the few ways in which we can learn about the properties of molecules that contain much energy.

Elementary Gas Reactions

We have seen how the energy requirements of bimolecular and unimolecular reactions can be met and must now turn to the problem of the cause of these requirements. The best approach is to consider first the facts about some of the simplest unimolecular reactions. In this context it is the formally simplest reactions that are important. Whether they are simple to investigate is another matter. In fact they are not. Our preoccupation is here in harmony with one of the main themes of modern kinetic research. Experimental complication is accepted for the sake of theoretical simplicity. The simplest major class of unimolecular reactions is that in which one bond only in a molecule is broken. For instance, ethyl benzene decomposes rapidly above 550°C:

$$C_6H_5CH_2CH_3 = C_6H_5CH_2 + CH_3 \quad (1)$$

Two free radicals are formed in all such reactions. Free radicals are very reactive. Unless special, rather complex, precautions are taken the free radicals undergo a mass of side reactions. No simple products are formed and the rate of reaction (1) cannot be found. The precautions lie outside the scope of this book. k is, however, simply given by the equation

$$\log k\,(\text{sec}^{-1}) = 14{\cdot}6 - (70{,}100/2{\cdot}303\,\boldsymbol{RT})$$

The rate constants of many similar reactions have been measured; some of them are listed in Table 3.1.

A normal C—C bond such as that joining the methyl and benzyl groups in ethyl benzene is about 1·5 Å long. A chemical bond is not of fixed length like an iron rod, it is better thought of as a spring which returns to a given length when not under stress.

TABLE 3.1

RADICAL DECOMPOSITIONS FOR WHICH THE ACTIVATION ENERGY IS BELIEVED TO BE EQUAL TO THE STRENGTH OF THE BOND BROKEN

Reaction	log A (sec^{-1})	E (kcal $mole^{-1}$)
$C_6H_5CH_2Br = C_6H_5CH_2 + Br$	13·0	50·5
$C_6H_5Br = C_6H_5 + Br$	13·3	70·9
$CH_2{:}CHCH_2Br = CH_2{:}CHCH_2 + Br$	12·7	47·5
$C_6H_5CH_2C_2H_5 = C_6H_5CH_2 + C_2H_5$	14·9	68·6
$C_6H_5NHCH_3 = C_6H_5NH + CH_3$	13·4	60·0
$CH_3NHNH_2 = CH_3NH + NH_2$	13·2	51·9
$(CH_3)_2NNH_2 = (CH_3)_2N + NH_2$	13·2	49·6

A stress can be applied either by the absorption of energy by radiation or by a collision. The two nuclei at either end of the elastic dumb-bell then vibrate so that the distance between them varies regularly. The movement of this atomic system is very similar to the simple harmonic motion found in the equivalent macroscopic assembly. The greater the energy content of the system the greater will be the amplitude of the vibrations. On the atomic scale this motion is represented by a "potential energy" diagram as in Fig. 3.5. The vibrations of atoms differ from those of classical harmonic systems in two ways. Firstly, they are quantized; but this difference is not important for our present purpose. Secondly, the potential energy curves depart from the parabolic form at high energy contents. The true curve is shown in the figure as a dotted line. When the bond becomes very short the repulsion between the atoms rises with increasing steepness. This is because the electron clouds that are not concerned in bond formation are being forced together and repel each other. On the other hand, as the bond gets longer the restoring force decreases. The figure shows that when the system has an energy content of 70 kcal $mole^{-1}$ a point on the potential energy curve is reached where there is no restoring force. Two atoms moving apart with this energy content will ultimately separate. This energy content

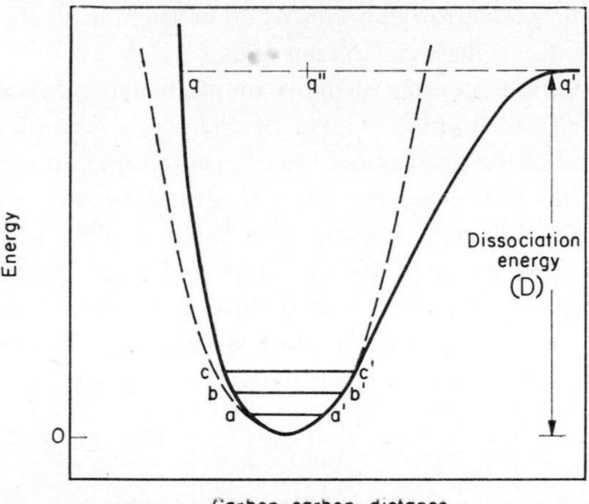

FIG. 3.5. A simplified potential energy diagram that represents the vibration of two atoms joined by a chemical bond. At absolute zero the atoms are at the rest point, the minimum of the curve. At higher temperatures the system can acquire more energy corresponding successively to the levels *a–a′*, *b–b′*, *c–c′*. The lettered points indicate the extreme dimensions between which oscillation occurs, they lie on a parabolic surface. The movement of the atoms corresponds to the movement of a ball rolling on this parabolic surface.

It can be seen that the real potential energy curve solid line deviates markedly from the ideal parabola (dotted line) at higher energies. Dissociation occurs when the level *qq′* is reached and there is no longer a barrier to dissociation.

is the bond dissociation energy. It is the heat of the reaction (1) and is represented in the figure by the height *D*. For a reaction that can be represented by the potential energy curve drawn with the solid line, the activation energy is equal to the heat of the reaction. Reactions in which a single bond is broken generally have activation energies that are equal to the heats of reaction. This is a very important generalization. It relates the kinetic property of activation energy to a property of a molecule at rest. The discovery of such relationships is a primary object of kinetics. Table 3.1 lists a number of reactions in which single bonds are broken. The activation energies of all of them are believed to be equal to the bond dissociation energies.

The activation energies of many unimolecular reactions cannot be predicted so simply. Those of the isomerization of cyclopropane and of the decomposition of cyclobutane, to mention two examples that have been considered, cannot be simply related to the strengths of their bonds. The same applies to the many decompositions of alkyl halides into hydrogen halide and the appropriate alkene. These activation energies cannot yet be predicted, and one can guess that no simple basis for prediction of the typical activation energies listed in Table 3.2 will ever be found.

TABLE 3.2

UNIMOLECULAR DECOMPOSITIONS IN WHICH A COMPLETE MOLECULE IS DIRECTLY ELIMINATED

Reaction	log A (sec^{-1})	E (kcal $mole^{-1}$)
$C_2H_5Cl = C_2H_4 + HCl$	14·2	59·5
$(CH_3)_2CHCl = C_3H_6 + HCl$	13·4	50·5
$(CH_3)_3CCl = C_4H_8 + HCl$	12·4	41·4
$C_2H_5Br = C_2H_4 + HBr$	13·4	53·9
$(CH_3)_2CHBr = C_3H_6 + HBr$	12·7	47·0
$(CH_3)_3CBr = C_4H_8 + HBr$	13·9	41·5
$C_2H_5I = C_2H_4 + HI$	13·4	50·0
$(CH_3)_2CHI = C_3H_6 + HI$	13·0	43·5
$(CH_3)_3CI = C_4H_8 + HI$	12·5	36·4

The A Factors of Unimolecular Reactions

Many of the reactions cited in Tables 3.1 and 3.2 have A factors between $10^{12 \cdot 5}$ and $10^{14 \cdot 5}$ sec^{-1}. There are also many reactions that have different A factors, but we can now recognize values around 10^{13} sec^{-1} as normal for reactions in which a free atom is liberated. Other values require special explanations. These rules were accepted early in the history of kinetics on the basis of evidence that is now known to be largely unreliable. The frequency of 10^{13} sec^{-1} can be understood in terms of the spring model that we have used to discuss reactions in which single bonds are broken.

Reference to Fig. 3.5 shows that an activated molecule is one which contains energy corresponding to the level qq'. At the moment that the molecule becomes activated the separation of the carbon atoms corresponds to a point between q and q'. Probably the separation is best represented by a point near the half-way mark q''. In order to find how rapidly the activated† molecule decomposes one must estimate the time taken for the atoms to separate from the point q'' to a point beyond q' where the potential energy curve is effectively flat. A rough answer can be obtained by considering motion within the bounds set by the dotted parabola. This parabola represents the condition of simple harmonic motion. Passage from one side of the parabola to the other and back takes one period of vibration. During reaction the separation of the atoms changes by a comparable amount. The time taken is also comparable. As the period of vibration is about 10^{-13} sec this will be the approximate lifetime of the activated molecule. The rate constant is the reciprocal of the lifetime; hence we get the frequency of 10^{13} sec^{-1}. Rigorous and elaborate theories of unimolecular reaction all establish a relation between the A factor of unimolecular reactions and the frequency of molecular vibrations. They differ in the way that this relation is approached. They allow for the complex nature of reacting molecules; whereas we have treated the system as diatomic.

The Activation Energies of Bimolecular Reactions

As yet we only understand the factors that influence the activation energies of the formally simplest bimolecular reactions. These are the atom transfer reactions such as

$$Br + CH_4 = HBr + CH_3 \qquad (2)$$

† It should be noted that the definition of an activated molecule is here slightly different from that adopted in the discussion on pressure effects.

Even in these instances our understanding is incomplete. Again, it has been necessary for the experimentalist to face considerable difficulties in order to obtain information on theoretically tractable systems. The bromine atoms for reaction are obtained by the photolysis or thermal decomposition of molecular bromine. The hydrogen abstraction reaction involves the rupture of the C—H bond in methane and the formation of the H—Br bond. The difference in the strengths of the bonds is the heat of the reaction

Strength of C—H bond in methane	104 kcal mole^{-1}
Strength of H—Br bond	87 kcal mole^{-1}
Heat of reaction	17 kcal mole^{-1}

The rate constant for reaction (2) is given by

$$\log k_1 \text{ (mole}^{-1}\text{ l. sec}^{-1}) = 10{\cdot}8 - (18{,}300/2{\cdot}303\, \boldsymbol{RT})$$

The connexion between the strengths of the bonds, the heat of the reaction and the activation energy is shown in the potential energy diagram of Fig. 3.6.

The diagram represents the transfer of the hydrogen atom from combination with a carbon atom to combination with a bromine atom. The carbon and bromine atoms are supposed to remain at a fixed distance apart while this occurs. The left-hand curve is a potential energy curve for the C—H bond. The right-hand curve is that for HBr; for this curve movement from left to right represents compression rather than extension of the bond.

Reaction is represented by a point moving on the surface of the curves between the two minima. In the course of its passage the point must surmount the energy barrier formed at the cross-over between the curves. The height of this barrier determines the activation energy.

Activation energies cannot be calculated from such simple curves. One reason is that the representation of the cross-over point as a sharp peak is unrealistic. It is always rounded. The extent to which this occurs cannot be reliably estimated. The normal alkanes, methane, ethane, propane and isobutane form an

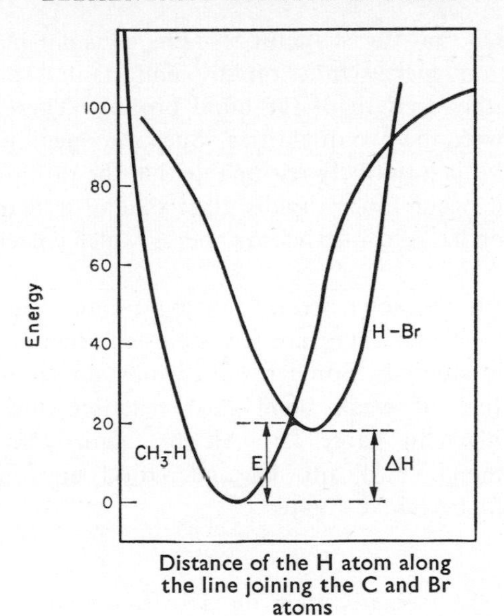

FIG. 3.6. Potential energy curves representing the reaction:

$$CH_4 + Br = CH_3 + HBr$$

Note that the activation energy is only slightly greater than the endothermicity of the reaction.

interesting series. The weakest C—H bond in each molecule is (kcal mole^{-1})

Me—H	Et—H	Pr^i—H	Bu^t—H
104	98	94·5	91·0

The rate constants at 100°C, *A* factors and activation energies for the reaction of bromine atoms with the most weakly bound hydrogen atoms are as follows:

	Me—H	Et—H	Pr^i—H	Bu^t—H
$\log k$ (mole^{-1} l. sec) 100°C	3·1	6·1	7·8	8·9
$\log A$ (mole^{-1} ml sec^{-1})	10·8	10·9	10·7	10·3
E (kcal mole^{-1})	18·3	13·4	10·1	7·5

It can be seen that the A factor remains constant in this series. The activation energy falls rapidly and its fall parallels the decrease in the strength of the bond broken. There is a linear relation between the two quantities. Such a connexion should not surprise us. It is intuitively reasonable that the rupture of a weak bond should occur more rapidly than that of a strong one. A rapid reaction has a low activation energy which corresponds to a low barrier.

Such simple relations are not always found even for simple hydrogen transfers. There are few series of reactions that have been reliably studied. Some results which do conform to the general pattern of weak bond—fast reaction—low activation energy are given in Table 3.3. All the factors that determine activation energies will not be understood until many more reactions have been investigated.

TABLE 3.3

REACTIONS THAT SHOW THE EXPECTED VARIATION OF ACTIVATION ENERGY WITH BOND STRENGTH

Reactions	$CH_3 + RH = CH_4 + R$ $CF_3 + RH = CF_3H + R$ $Cl + RH = HCl + R$		
Reactant RH	CH_3	CF_3	Cl
	Activation energy (kcal mole^{-1})		
CH_4	12·8	10·3	3·9
C_2H_6	10·4	7·5	1·0
n-C_4H_{10}	8·3	5·1	0·7
iso-C_4H_{10}	7·5	4·7	0·7

Proton Transfer Reactions

It might be thought that reactions that involve the transfer of a proton would be easier to understand than the transfer of complete hydrogen atoms. This is not so because proton transfers are studied in solution. The charged reactants and products are known to be heavily solvated but little progress has been made in the quantitative description of their condition. A proton transfer reaction of historical importance is the decomposition of nitramide

$$NH_2NO_2 = N_2O + H_2O$$

In acid or neutral solution the reaction is slow, but it is rapid when a base is present. If the base, which is defined as a species with the capacity to accept a proton, is a hydroxyl ion then the mechanism of the reaction is

$$NH_2NO_2 + OH^- = NHNO_2^- + H_2O$$
$$NHNO_2^- = N_2O + OH^-$$

The hydroxyl ion which is consumed in the first step is regenerated in the second so that its concentration does not vary during the reaction. The reaction is followed by measurement of the rate of formation of nitrous oxide which is given by

$$R_{N_2O} = k[NH_2NO_2] = k_0[NH_2NO_2] + k_c[OH^-][NH_2NO_2]$$

k is a first-order rate constant, but its value in any run depends upon the hydroxyl ion concentration. A similar expression is found if the hydroxyl ion is replaced by another base such as aniline, when

$$k = k_0 + k_c'[NH_2NO_2][C_6H_5NH_2]$$

In either case, several runs are performed and k is plotted against the concentration of base. A straight line of intercept k_0 and slope k_c' is obtained.

Hydroxyl ion and aniline increase the rate of the decomposition but themselves emerge unchanged. They satisfy the classical definition of a catalyst. This branch of chemistry is known as acid–base catalysis; many of the reactions, such as the mutarotation of glucose, are catalysed by both acids and bases. The nomenclature is, however, unfortunate because it tends to suggest that special kinetic principles should apply. This is not so; the rate

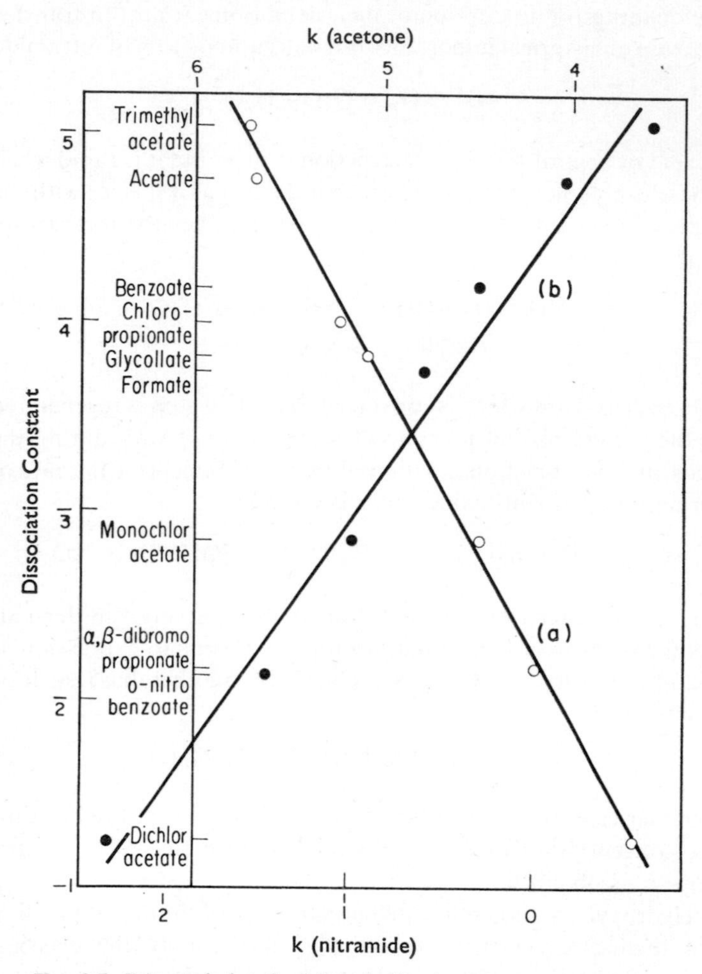

FIG. 3.7. Brönsted plots for (*a*) the iodination of acetone in aqueous solution at 25°C catalysed by carboxylic acids, (*b*) the decomposition of nitramide in aqueous solution at 25°C catalysed by the basic anions of carboxylic acids,

constants of the catalysed reactions should be treated in the same way as those of any homogeneous system. The only peculiarity is the regeneration of one of the reactants.

The decomposition of nitramide was studied by Brönsted and his co-workers. They found it to be catalysed by many classes of base, including uncharged amines, negatively charged anions and positively charged metal complexes. In each case, first-order rate constants were measured for several concentrations of base and the second-order constants, k_c, which are known as the catalytic coefficients, were derived. The rate constant, k_0, can be regarded as a product of the catalytic coefficient of water multiplied by its effective concentration.

The normal procedure, when studying catalysed reactions, is to measure the catalytic coefficients of several bases of similar chemical structure. Brönsted found that the coefficients are then related to the basicity of catalysts as measured by their dissociation constants in water. This Brönsted relation, as it is called, is shown both for an acid and a base catalysed reaction in Fig. 3.7. The results plotted are for series of acids and of bases of very similar structure. The inclusion of acids and bases of different structure would have introduced some scatter into the points but for most reactions the relations would still have been easily apparent. These relations are usually written in either the form

$$k_c = GK^{\alpha} \quad \text{or} \quad \log k_c = \alpha \log K + \log G$$

where K is the dissociation constant of the catalyst and α is the Brönsted coefficient which always lies between 0 and 1. The existence of this relation should not surprise the reader. The rate determining step in the decomposition of nitramide catalysed by aniline is

$$NH_2NO_2 + C_6H_5NH_2 = NHNO_2^- + C_6H_5NH_3^+$$

It should be expected that the greater the attraction of the aniline for the proton as measured by the dissociation constant, the greater will be its catalytic power. The connexion is clearly perceived if the equilibrium constant is written for the reaction:

$$H_3O^+ + C_6H_5NH_2 = H_2O + C_6H_5NH_3^+$$

Similarly a reaction in which the substrate accepts a proton is most strongly catalysed by substances that readily donate protons. These substances are the acids that have large dissociation constants.

Logarithmic relations between rate constants are frequently found when series of reactions of similar compounds are studied. Another classical example is provided by the reactions of substituted phenyl derivatives which were particularly studied by Hammett. He found that the rate constants for the reactions of benzoyl chloride with series of anilines of structure, $XC_6H_4NH_2$, are related to the dissociation constants of the corresponding benzoic acids, XC_6H_4COOH (X is either an *m* or *p* substituent). The logarithms of the rate constants when plotted against the logarithms of the dissociation constants yield a straight line. Similar straight lines are found if either set of logarithms is plotted against the logarithms of the rate constants for the alkaline hydrolysis of substituted ethyl benzoates. As yet the slopes of these lines cannot be predicted.

CHAPTER 4

THE KINETICS OF COMPLEX REACTIONS

FEW reactions that can be studied by direct observation of reactants and products are elementary. Most systems involve a succession of two or three steps. One of these steps is usually much slower than the others and is, therefore, rate-determining. The consequences of this fact and the way it is used to determine mechanisms are discussed in the next chapter. There are also systems in which the overall reaction can only be satisfactorily described in terms of many successive steps. These systems have been intensively studied because some are of great economic importance. They also introduce novel kinetic principles; it is for this reason that they are discussed here.

Hydrogen and chlorine can remain together in the dark at room temperature for an indefinite time. The gases may be contained unchanged in a glass bulb with two electrodes either with a short gap between them or joined by a fine filament. If a spark is passed or the filament momentarily heated the whole mixture will react. Very probably the reaction will be so violent as to shatter the bulb. Intense light will also cause an explosion. Under weak illumination the reaction is more moderate. The quantum yield is still very high. Its precise measurement is likely to be difficult because of the irreproducibility of the experiments but the value will lie between 10^3 and 10^7 if the gases are reasonably pure. These facts and many more were known in the opening years of this century when the study of the reaction occupied the masters of gas kinetics. Probably no other gas reaction has been so exhaustively studied. In order to explain the observations, Nernst (1913) suggested an atomic *chain mechanism* which in broad outline has since been universally accepted. He suggested that reaction began by the dissociation of chlorine molecules

$$Cl_2 = Cl + Cl$$

The atoms reacted with hydrogen molecules, yielding hydrogen atoms

$$Cl + H_2 = HCl + H$$

which subsequently reacted with chlorine molecules

$$H + Cl_2 = HCl + H$$

The cycle of these two stages repeated many times forms the chain of reactions from which the mechanism gets its name. The chains are ended by the reaction of hydrogen or chlorine atoms to form unreactive products. Two chlorine atoms may combine; this is the reverse of the initiation reaction. Alternatively a chlorine atom may combine with a hydrogen atom or a hydrogen atom may stick to the wall of the reaction vessel where it is likely to combine readily with a second hydrogen. The exact determination of the way in which the atoms are removed is very difficult. It will vary according to the nature of any minute quantities of impurities that are present. No kinetic order can be ascribed to the overall reaction.

Nernst's proposal was revolutionary because for the first time definite evidence was forthcoming that free atoms were involved in a chemical process. Similar mechanisms describe the reaction of chlorine with paraffins and many other hydrogen-containing compounds. These processes involve organic free radicals in place of hydrogen atoms. They are important industrially for the synthesis of chlorinated ethanes and ethylenes which are used as grease solvents, particularly in dry cleaning.

The reaction of hydrogen with chlorine is a simple example of a reaction that might be supposed to be molecular, as is the reaction of hydrogen with iodine. The study of their kinetics immediately shows that the mechanisms are different. For instance, visible light does not alter the rate of the reaction with iodine. The difference between a molecular reaction and a free radical chain reaction is not always so obvious. This partially accounts for the delay in the recognition of processes of the latter type. Two examples are the decompositions of ethane and acetaldehyde. They can be written:

$$C_2H_6 = C_2H_4 + H_2$$

and

$$CH_3CHO = CH_4 + CO$$

The equations are to within one per cent full descriptions of the changes. Furthermore the rate of formation of ethylene is given by

$$R_{C_2H_4} = k[C_2H_6]$$

which would be expected for a unimolecular process. The chain nature of the reaction is, however, revealed by the effect of small concentrations of additives. Some, such as ethers, markedly accelerate the reaction; others, such as nitric oxide, retard it. The ethers should not be thought of as catalysts for they are consumed in the reaction. They are initiators. Their function can be shown by considering the steps that constitute the chain:

$$C_2H_6 = 2CH_3 \qquad (1)$$

$$CH_3 + C_2H_6 = CH_4 + C_2H_5 \qquad (2)$$

$$C_2H_5 = C_2H_4 + H \qquad (3)$$

$$H + C_2H_6 = H_2 + C_2H_5 \qquad (4)$$

$$H + C_2H_5 = C_2H_6 \qquad (5)$$

The rate constant of reaction (1) is low. Little ethane decomposes in this way and few ethyl radicals are formed by reaction (2). All but a fraction of one per cent of the ethane is removed by reaction (4) which is followed by reaction (3). Conventionally (1) is called the initiation reaction; (3) and (4) are the chain-carrying reactions; (5) is the chain termination. This sequence of initiation, chain-carrying and termination is characteristic of all chain reactions. The concept of chain length is often employed in the discussion of chain reactions. For ethane it can be defined as:

$$\text{Chain length} = \frac{\text{Number of molecules removed by reaction (4)}}{\text{Number of molecules removed by reaction (1)}}$$

Precise definition is not always simple, but the idea behind the concept is clear. The chain length is intended to measure the number of times that the chain-carrying cycle repeats for each introduction of an active radical by initiation. The overall rate could be increased by:

(i) Increasing the rate of initiation.
(ii) Increasing the rate of development of the chain. (This means that either reaction (3) or (4)—whichever is the slower—must be accelerated.)
(iii) Decreasing the rate of termination.

The decomposition of ethane is accelerated by ethers because they increase the rate of initiation. Reactions (3), (4) and (5) are all elementary, so that their rate constants cannot be affected by additives. It has been supposed that nitric oxide decreases the rate of the reaction because it reacts rapidly with free radicals. It presumably reduces the concentrations of the chain-carrying radicals and may provide an alternative termination reaction. Its precise mode of action is still a matter of debate.

The decomposition of ethane is not accelerated by ultraviolet light with wavelength greater than 2500 Å because none of the species present absorbs the radiation. On the other hand, light markedly increases the rate of decomposition of acetaldehyde. The quantum yield varies from low values up to more than one hundred depending on the conditions; a high temperature particularly favours long chains. The photolysis which occurs yields free radicals that initiate chains. The quantum yield is a measure of chain length, if each photolytic free radical initiates a chain.

Chain reactions of great industrial importance are vinyl polymerizations. They yield such products as

"Perspex" (polymethyl methacrylate)	$(CH_2{=}C(CH_3)COOCH_3)_n$
polyvinyl chloride	$(CH_2{=}CHCl)_n$
polystyrene	$(CH_2{=}CH{-}C_6H_5)_n$

The polymer consists of a long molecule formed by opening the double bonds in the monomer molecules so that they can join up. This is shown for a generalized vinyl monomer,

$$CH_2{=}CHX$$

$$CH_2{=}CHX \quad CH_2{=}CHX \quad CH_2{=}CHX \quad CH_2{=}CHX$$

$$-CH_2-CHX-CH_2-CHX-CH_2-CHX-CH_2-CHX-$$

As with other chain reactions, the overall process can be described in terms of initiation, chain-carrying steps and termination. Since polymers are of high molecular weight and involatile, polymerizations are generally performed in the liquid phase, but it has been shown that the process can occur in a gas. Some polymerizations can be satisfactorily initiated by heating the monomer. Ultraviolet light will initiate others. It is, however, usual to introduce an initiator. This is a compound, such as di-t-butyl peroxide, $(CH_3)_3COOC(CH_3)_3$, that decomposes into free radicals at a low temperature (below 150°C). Initiation can thus be readily controlled.

The chain-carrying steps simply involve the addition of successive monomer units to the growing polymer radical:

$$R- + CH_2{=}CHX = RCH_2-CHX-$$

Despite much investigation the precise nature of the termination steps is not known. Kinetic studies show that they require the interaction of two polymer radicals, but the high molecular weights make it impossible to identify products directly.

For polymers the term "chain length" has an obvious meaning. It is simply the average number of monomer units to be found in the polymer products. Values of a hundred to a thousand are common.

Polymerization can be stopped by inhibitors which are usually hydroquinones or aromatic diamines. Many of these substances, which must react rapidly with free radicals, have been developed. They are important because without their aid it would be difficult to store monomers. Similar compounds are used to stop other free radical chain processes, such as the oxidations that cause the perishing of rubber and the formation of gums in petrol.

The chain reactions discussed above are alike in that the reaction of a free radical in a chain-carrying step results in the formation of only one new free radical. This need not be so. The hydrogen–oxygen reaction involves reactions such as

$$O + H_2 = O + OH$$

in which one free atom yields a free atom and a radical, as well as those, such as

$$OH + H_2 = H_2O + H$$

in which a free radical is simply replaced by a free atom so that the number of unsatisfied valencies remains constant. When two active molecules are formed in place of one, the reaction accelerates. It is what is known as a branching chain. Under suitable conditions the chain will branch many times, so that the reaction becomes uncontrollably fast. The mixture then explodes. The cause of the explosion is different from that with hydrogen and chlorine, which occurs when the reaction accelerates with a rise in temperature. The rise is caused by heat produced in the reaction that cannot be removed as rapidly as it is generated. Branching chain reactions will also often accelerate more than the chemical mechanism would imply, because the heat generated is not rapidly dissipated. The exact prediction of the progress of such reactions is very difficult. Even when all the chemical information is available, the problem of solving the heat transfer equations remains. Complete solutions have been found only for a few special cases such as reactions in spherical vessels.

CHAPTER 5

KINETICS AND MECHANISM

ONE of the most striking advances in organic chemistry over the last forty years has been the great increase in knowledge about the mechanism of reactions. The development started with the acceptance of the electronic theory of valency without which understanding was impossible. The only kinetic information that was used was the figures for relative yields of products. An example is the relative yields of *ortho*, *meta* and *para* substituted benzenes formed on, say, nitration of aniline. Forty years ago, intensive kinetic studies of several basic organic reactions were begun with the avowed object of elucidating their detailed mechanisms. Today the tactics of determining a mechanism by kinetic studies are understood but there are still many systems that have not been explored. In this chapter we shall consider two classes of reaction that are of historic importance and which are good examples of the kinetic approach. They are the halogenation of ketones and substitution of a saturated carbon atom in an lkayl halide.

The Halogenation of Ketones

The halogenation of the simplest ketone, acetone, is represented by the equation

$$Br_2 + CH_3COCH_3 = HBr + CH_3COCH_2Br$$

when the halogen is bromine. If the equation described an elementary reaction, the rate of formation of the brominated ketone would be given by

$$R_{CH_3COCH_2Br} = k[CH_3COCH_3][Br_2]$$

The reaction can be studied in aqueous solution by following the rate of disappearance of bromine; usually iodometric titration is employed, but other methods are sometimes convenient. If the acetone is in excess, its concentration will not effectively alter during a run. It is then found that

$$R_{CH_3COCH_2Br} = k[CH_3COCH_3]$$

provided that conditions in the system do not otherwise alter. This rate equation is incompatible with the elementary mechanism because the concentration of bromine is not involved. A few experiments reveal that the rate of the reaction depends upon the hydrogen ion concentration, so that satisfactory runs can only be performed in buffer solutions. Series of runs can then be performed in which the hydrogen ion concentration remains constant, but in which the concentrations of the constituents of the buffers are varied. Such experiments quickly show that the reaction is generally catalysed by both the acids and bases which constitute the buffers. The rate of formation of bromoacetone can then be written:

$$R_{CH_3COCH_2Br} = k_0[CH_3COCH_3]+k_a[CH_3COCH_3][A] \\ +k_b[CH_3COCH_3][B]$$

where [A] and [B] are the concentrations of acid and base present, as defined by the Brönsted definition as substances able to donate or receive a proton. Conditions can be adjusted so that either the second or third term is very small. The experimental facts then indicate that the reaction is complex and that the rate-determining step involves the transfer of a proton. Lapworth (1904) thought that the ketone must always be converted into its enol form before it would react with the halogen.

$$CH_3COCH_3 \rightarrow CH_2\!:\!C(OH)CH_3$$

Further study showed that this need not occur, and the following mechanisms have since been accepted as reasonable, if not perfect, descriptions for halogenations of ketones.

Basic catalysis

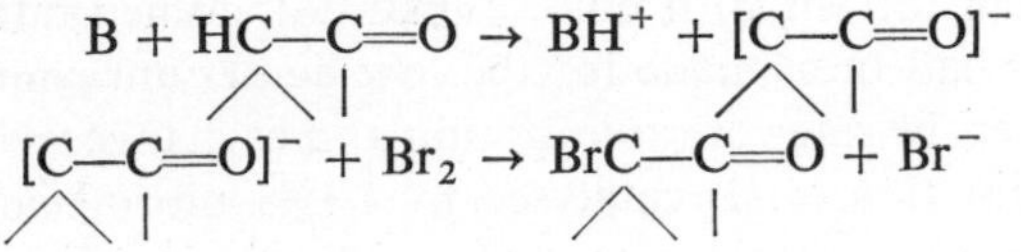

$$\mathrm{B + HC{-}C{=}O \rightarrow BH^+ + [C{-}C{=}O]^-}$$

$$\mathrm{[C{-}C{=}O]^- + Br_2 \rightarrow BrC{-}C{=}O + Br^-}$$

Acid catalysis

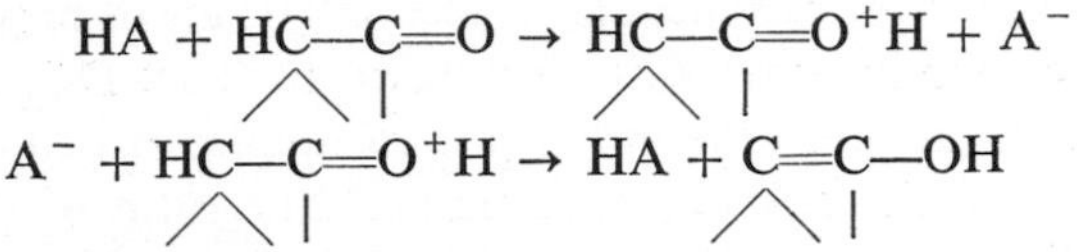

$$\mathrm{HA + HC{-}C{=}O \rightarrow HC{-}C{=}O^+H + A^-}$$

$$\mathrm{A^- + HC{-}C{=}O^+H \rightarrow HA + C{=}C{-}OH}$$

The enol then reacts rapidly with the halogen.

There are two other reactions whose mechanisms are closely related to those for halogenation. They are the racemization of ketones with asymmetrical α-carbon atoms and the isotopic exchange of the α-hydrogen atoms with heavy water. In the first case it is clear that the tetrahedral conformation of the α-carbon atom cannot be retained in either the anion or the enol. The easy reversibility of all of the steps ensures that isotopic exchange must occur as rapidly as the anion or enol is formed. The confirmation of these predictions is the most striking evidence for the two mechanisms.

The rate constant for the bromination of

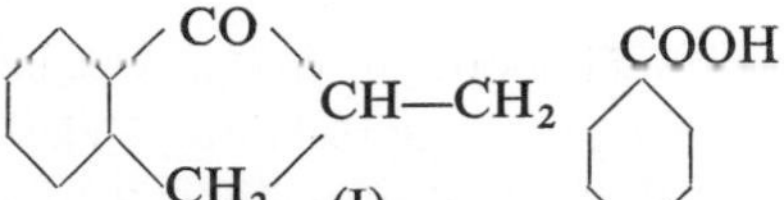

CO, CH—CH_2, CH_2, COOH (I)

in a 0·05 M solution of sodium acetate in 16 N acetic acid is 0·0471 hr^{-1}. The rate constant for the racemization of the *d*-ketone is 0·0438 hr^{-1}.

The racemization of

Et, Me >CHCOPh (II)

was studied simultaneously with its isotopic exchange in a D_2O-dioxan solvent with OD^- as catalyst. When rather a large correction had been made for the low deuterium content of the heavy water the rates of racemization and exchange were equal.

The iodination of (I) catalysed by 1·19 N nitric acid in glacial acetic acid has a rate constant of 0·0298 min^{-1}. This is, to well within the probable experimental error, equal to the rate constant of 0·0296 min^{-1} for racemization. The initial rate of bromination of light acetone in heavy water catalysed by 0·3 N DCl has been measured; so also has the initial rate of uptake of deuterium by the acetone in the absence of bromine. The ratio of the rates, found in three determinations, were 1·05, 0·90, and 0·95.

The observations on the rates can be summarized (following Ingold):

For basic catalysis

Halogenation	in	light solvent	are proved equal
Racemization	in	light solvent	
		heavy solvent	are proved equal
H exchange	in	heavy solvent	

For acid catalysis

Racemization	in	light solvent	are proved equal
Halogenation	in	light solvent	
		heavy solvent	are proved equal
H exchange	in	heavy solvent	

Both sets of results point strongly to the existence of a single type of intermediate for each class of reaction. Unless one can suppose that these intermediates exist, it is difficult to see why the rates should be the same. The postulated anions and enols appear to meet all the requirements of our present state of knowledge. It should not, however, be overlooked that there are no positive proofs of mechanisms, but only degrees of certainty. In the present instance, despite all the work that has been done, the mechanisms of these interrelated reactions still contain many points of interest that are not understood and are the subjects of continued investigation.

Substitution on a Saturated Carbon Atom in an Alkyl Halide

One of the most common reactions of aliphatic chemistry is nucleophilic substitution at a saturated carbon atom. An example is the Finkelstein reaction:

$$I^- + RCl = RI + Cl^-$$

Others are the hydrolysis or alcoholysis of alkyl halides:

$$H_2O + RCl = ROH + RCl$$

$$OEt^- + RCl = ROEt + Cl^-$$

Even before the development of the electronic theory of valency, several mechanisms were suggested which have since been shown to contain a measure of truth. They were not widely accepted because there was no direct evidence for them. The evidence only came when full kinetic studies of the reactions were made.

Hughes, Ingold and their collaborators have been chiefly responsible for the study of the hydrolysis of alkyl halides. They compared the rates of reaction of different bromides under identical conditions. The same solvent, 80 per cent ethanol and 20 per cent water, was suitable for all the compounds. The reactions proceeded at such different rates that they could not all be conveniently studied at 55°C which was the temperature selected for comparisons. Methyl and ethyl bromides were studied at 55°C. For trivial reasons, isopropyl bromide was studied at temperatures above and below 55°C and the required value found by interpolation, using the Arrhenius equation. t-Butyl bromide was studied at 0·10°C, 15·05°C and 24·95°C and the rate constant at 55°C found by extrapolation using the Arrhenius equation. The experimental arrangements were similar for all the bromides. The extent of the reaction was estimated at convenient intervals by titration of the reaction mixture with standard acid, alkali or silver nitrate, whichever was appropriate. At low temperatures, a considerable quantity of reaction mixture was made up and samples withdrawn for analysis. At high temperatures, the more volatile reactants would have been lost while the sampling was taking place. The reaction mixtures were, therefore, sealed in ampoules which were rapidly chilled and broken open when they were removed from the thermostat at the end of the selected period. The higher bromides lose hydrogen bromide as well as hydrolyse. The extent of this side reaction was determined by analysis for olefins. A correction has accordingly been applied to the rate constants where it is necessary.

The runs carried out in the pure solvent gave first-order rate constants as shown in Table 5.1. The rates of reaction of methyl and ethyl bromide were too low to be detected. Runs were then performed in dilute solutions of sodium hydroxide. Second-order rate constants were found with methyl and ethyl bromide; the rates were first order with respect to both the bromide and the hydroxide ion. For convenience values are given in Table 5.1 for

TABLE 5.1

THE HYDROLYSIS OF ALKYL BROMIDES

Hydrolysis in 80% ethanol, 20% water by volume at 55°C.
All rate constants multiplied by 10^5

	MeBr	EtBr	Pr^iBr	Bu^tBr
First-order rate constants in pure solvent (sec^{-1})	—	—	0·24	1010
First-order rate constant for removal of alkyl bromide in 0·01 N Na OH (sec^{-1})	21·4	1·7	0·29	1010
Second-order rate constant for reaction of hydroxyl ion with alkyl bromide ($mole^{-1} l. sec^{-1}$)	2140	170	4·7	—

first-order rate constants obtained when the concentration of sodium hydroxide was steady at 0·01 N. It can be seen that the rate of hydrolysis of t-butyl bromide does not depend upon the concentration of hydroxide. The rates of reaction of methyl bromide and ethyl bromide are now large. These compounds evidently hydrolyse only by second-order mechanisms. Their second-order rate constants are found by dividing the first-order constants by the concentration of hydroxide (0·01 N) which gives the values shown in Table 5.1.

Isopropyl bromide is more complicated. The first-order rate constant in 0·01 N hydroxide is 0·05 sec^{-1} greater than in pure solvent. Hydrolysis must occur by two mechanisms simultaneously. The second-order rate constant of about 5 $mole^{-1}$ l. sec^{-1} could be found by simple division but the determination would be inaccurate as it would depend upon the measurement of a small difference in two relatively large quantities. A better value was found from runs with approximately 0·8 N hydroxide; a correction of only 3 per cent was then required to allow for the reaction in the pure solvent.

The relation between the results can best be seen from a diagram in which the logarithms of the rate constants in 0·01 sodium hydroxide solution are plotted. Many observations indicate that the substitution of alkyl groups largely affects the rates of organic reactions through the variation in the ability of the groups to

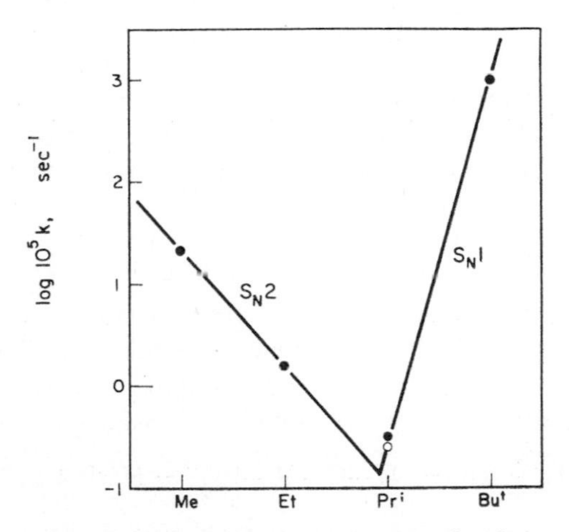

FIG. 5.1. Plot of the first-order rate constants for the reaction of alkyl bromides in aqueous ethanol at 55°C. The full points are experimental determinations; the open point for isopropyl bromide is the contribution of the rate found in pure solvent corresponding to the direct dissociation mechanism.

release electrons. An extreme manifestation of this effect is the variation in energies required to form carbonium ions from hydrocarbons in the gas phase. The heats of reaction are as follows:

$$\begin{array}{rcll} CH_4 &=& CH_3^+ + H + e^- & 336\ \text{kcal mole}^{-1} \\ C_2H_6 &=& C_2H_5^+ + H + e^- & 298 \\ C_3H_8 &=& (CH_3)_2CH^+ + H + e^- & 267 \\ (CH_3)_3CH &=& (CH_3)_3C^+ + H + e^- & 250 \end{array}$$

Most of the decrease can be attributed to the stabilization of the positive charge on the carbon atom by the electron-releasing effect of each methyl substituent. Since carbonium ions are possible intermediates in the hydrolysis of the halides, it is clearly worth while to see what predictions can be made along these lines. In the first place if reaction takes place by the mechanism

$$RBr = R^+ + Br^-$$

followed by

$$R^+ + OH^- = ROH$$

then the rate constants should increase with the ease of formation of the carbonium ion. This increase is observed for $Bu^t \gg Pr^i \gg Et$ and Me. Furthermore if the rate of ionization is rate-determining, the hydrolysis should be first-order with respect to the alkyl halide and independent of the concentration of hydroxyl ion.

The much smaller variation of the second-order constants cannot be simply predicted. A detailed quantitative study is even needed to predict the direction of the change. The difficulty arises because the mechanism of the second-order reaction involves the activated complex shown

$$OH^- + RBr = (H\text{—}O\text{—}R\text{—}Br)^- = HOR + Br^-$$

Substitution on R can only slightly alter the distribution of charge and hence the stability of the complex. n-Propyl, n-butyl and n-amyl bromides have second-order rate constants only slightly less than that of ethyl bromide.

CHAPTER 6

THE KINETICS OF HETEROGENEOUS CATALYSIS

THE term "heterogeneous catalysis" can be properly applied to all reactions that occur more rapidly in the presence of a surface than in a homogeneous system. Mild catalysis is very common. Many homogeneous gas reactions can only be studied if proper precautions are taken to eliminate concurrent heterogeneous reactions on the glass or quartz surface of the reaction vessel. This is usually done by poisoning the surface by long contact with reactants and products. The term is, however, more frequently used to describe enormous increases in the rate of such reactions as that of nitrogen with hydrogen brought about by the presence of iron oxide in the Haber process. Similar catalysis is observed with some reactions of liquids and in the presence of solvents, but at the present time the catalysis of reactions between gases is of greatest industrial importance. Some examples are the cracking of petroleum on aluminium oxide, the combination of oxygen and sulphur dioxide on platinized asbestos, and the isomerization of hydrocarbons on platinum. A catalyst can, of course, only facilitate the progress of reactants towards an equilibrium mixture. It can, therefore, be said to accelerate reactions that are already occurring. The difference in behaviour in the presence and absence of catalyst is often so great that it is difficult to remember this fact. Sometimes the same reactants will yield different isolable products with different catalysts. Thus carbon monoxide and hydrogen form hydrocarbons by the Fischer–Tropsch reaction over cobalt, thorium, and magnesium oxides on kieselguhr, but methanol over zinc oxide. Even more striking are the reactions of ethanol. Over alumina above 300°C the products are ethylene and water; between 260° and 300°C they are diethyl ether and water; over copper between 200° and 300°C they are acetaldehyde and hydrogen.

In 1833 Michael Faraday wrote of *The Power of Metals and Other Solids to Induce the Combination of Gaseous Bodies.* Since that date there has been continuous investigation of heterogeneous catalysis. The tempo has increased rapidly with the growth of the petrochemical industry. The art is well developed as a result of intensive series of trial-and-error experiments. For example, over thirty years Mittasch and his co-workers tested thousands of formulations as catalysts for the ammonia synthesis. It is now often possible to select the right type of compound for a particular purpose. Nickel and iron are known to be effective in the hydrogenations of organic compounds ($C_2H_4+H_2 = C_2H_6$), but useless for dehydrations of alcohols ($C_2H_5OH = C_2H_4+H_2O$) which are accelerated by alumina. Recently science has begun to catch up by the application of the same principles that have been fruitful in the study of homogeneous reactions. Investigators now try to study the simplest and best-defined systems. Again the experimental difficulties are great.

For synthetic work a catalyst is most simply packed in the middle of a tube. Its temperature can be controlled by a simple furnace. The gases flow in at one end and react. As they pass from the catalyst chamber the reaction mixture will be effectively frozen. Most catalysts slowly lose their effectiveness, frequently because of poisoning by impurities in the reactants. When the catalyst is plugged in the middle of a tube, the reaction must be stopped for its replacement or regeneration. For large-scale processes continuous operation is achieved either by having a moving bed of catalyst or by fluidizing. Fluidization depends upon the fact that small particles in a gas stream behave as a liquid which can be circulated through regeneration chambers by pumps. In the laboratory much of the most significant work has been done on metal films formed on the surface of vessels by the evaporation of hot wires in a high vacuum. This is one of the best ways of obtaining a really clean surface.

If the concentrations of reactants is kept constant, then the rate of a catalysed reaction, measured by the rate of formation of products is given by:

$$\text{Rate} = B \exp(-E/\boldsymbol{R}T)$$

This generally holds over the range of temperature accessible to experiment. The symbol B is used in place of the more familiar A to emphasize that it is a different kind of parameter. It is roughly proportional to the surface area of the catalyst. E is readily reproducible for a particular reactant-catalyst pair. It is invariably lower than the activation energy of the homogeneous reaction. Thus for the decomposition of hydrogen iodide, the activation energy of the homogeneous reaction is 44 kcal mole^{-1} but E is only 25 and 14 kcal mole^{-1} on gold and platinum respectively. This lowering is sometimes represented by potential energy curves with barriers of different heights. Such diagrams should be studied with caution because of the difficulty of selecting a suitable reaction co-ordinate to plot on the abscissa. It is impossible to calculate values of E from first principles.

The Mechanism of Heterogeneous Catalysis

Catalysts were once thought to be active because the molecules of reactant absorbed on their surfaces were at higher effective concentrations than those in the gas. This view has been rejected because, for some reactions, many solids have been found that physically adsorb the reactants. The adsorption is as strong as with the catalyst, but most such solids are inactive. Chemisorption is now accepted as a prerequisite for activity.

One of the reactions for which the mechanism of chemisorption is best understood is the hydrogenation of ethylene, particularly on a nickel catalyst. Ethylene could be adsorbed either by the opening of the double bond (I) or by the separation of a hydrogen atom (II).

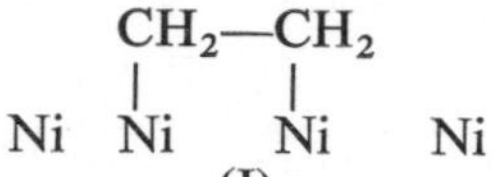

```
                                                CH2
                                                ||
        CH2—CH2                                 CH  H
        |     |                                 |   |
    Ni  Ni    Ni    Ni                      Ni  Ni  Ni  Ni
          (I)                                    (II)
```

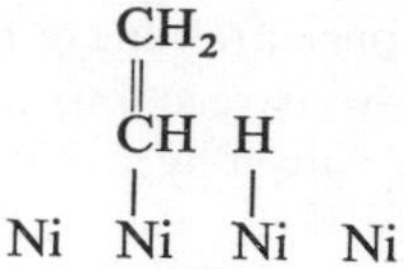

The second arrangement is unlikely as no isotopic mixing occurs when C_2H_4 and C_2D_4 are chemisorbed together.
(I) can react with physically adsorbed hydrogen

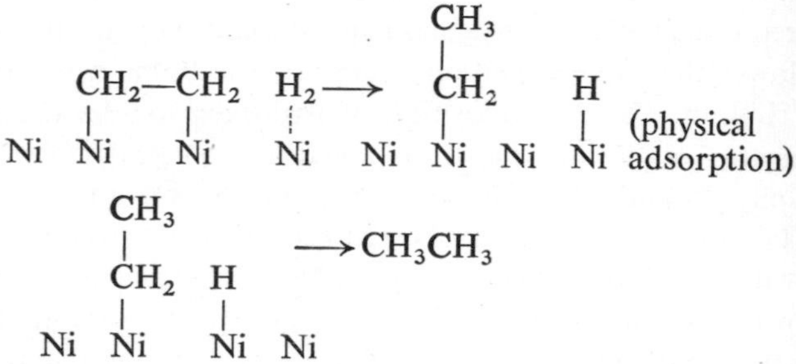

```
                                      CH3
                                      |
      CH2—CH2    H2 ──→               CH2        H
      |     |    ¦                    |          |  (physical
  Ni  Ni    Ni   Ni             Ni    Ni    Ni   Ni adsorption)

        CH3
        |                  ──→ CH3CH3
        CH2    H
        |      |
    Ni  Ni     Ni  Ni
```

This mechanism is believed to describe one of the ways in which the reaction occurs. There is good evidence that other mechanisms also contribute.

Although the above scheme suggests the way in which the heterogeneous reaction occurs it does not tell us why nickel is a particularly effective catalyst. Two explanations suggest themselves. First nickel may form particularly suitable bonds with the chemisorbed reactants. Second the chemisorbed molecules may be placed on the surface so that they can readily react.

The strength of the bonds formed by ethylene can be deduced from the amount of heat released when one molecule of ethylene is adsorbed on nickel. The amount is smaller than for metals such as iron, chromium and tantalum which are worse catalysts. Rhodium is a better catalyst and has a lower heat of adsorption. These facts indicate that active catalysts tend to form weak bonds with the substrate. The relative positions of molecules on the surface cannot be studied so readily. Even a thin evaporated film is crystalline and the arrangement of atoms is different on the different crystal faces. Fortunately, methods have been found of depositing nickel films in which a selected face is exposed. In this way, it has been shown that nickel atoms 3·5 Å apart (one common distance) are more active than those 2·5 Å apart (another common distance).

The geometrical interpretation of catalytic activity is not as straightforward as the above facts would suggest. Minute quantities of some poisons will greatly reduce the activity of a catalyst without having a similar effect on its powers of adsorption. Thus a small quantity of mercury vapour will reduce the rate of hydrogenation of ethylene on copper to 0·5 per cent although 80 per cent of the adsorption power for ethylene and 5 per cent for that of hydrogen are retained. This together with numerous other facts on poisoning suggest that there are certain active sites on the catalyst surface that contribute an amount of reaction out of all proportion to their numbers.

On the other hand, there are substances which although they are themselves inactive can enormously increase the activity of catalysts to which they are added. They are known as promoters. An example is the promotion of the iron oxide catalyst for the ammonia synthesis by the addition of alumina and some potash. The mechanism of promoter action is not understood. Very likely the action is peculiar to each catalyst if not each promoter. It seems that the primary effect of the promoter is to improve and maintain the microscopic structure of the surface of the catalyst.

PROBLEMS

CHAPTER 1

1. The decomposition of di-t-butyl peroxide can be represented by the overall equation

$$(CH_3)_3COOC(CH_3)_3 = 2CH_3COCH_3 + C_2H_6$$

The following measurements were obtained from a study of the decomposition of the peroxide at 155°C.

Time (sec)	0	180	360	540	720	900
Pressure (mm)	173·5	193·4	211·3	228·6	244·4	259·2

Calculate the first-order rate constant.

2. In a first-order reaction 30 per cent of the reactant has disappeared after 500 sec. What is the rate constant of the reaction? How long must elapse before 60 per cent has disappeared?

3. The following observations were made on the isomerization of 1,2-dimethylcyclobutene to 2,3-dimethylbuta-1,3-diene at 170·7°C.

Time (sec)	1200	4200	6120	9000
% cyclobutene remaining	86·2	59·1	46·5	32·6

Determine the first-order rate constant by graphical means.

4. The half-life of N^{13} is 10·1 min. What is the rate constant of the decay reaction?

5. The half-life of C^{14} which decays radioactively to N^{14} and an electron is 5760 years. What proportion of a sample of C^{14} will remain after 1965 years?

6. One gramme of radium is placed in an evacuated tube of 10 ml at 25°C. Each radium nucleus that decomposes yields 4 helium atoms. After two years the pressure of helium in the tube is 28 mm. What is the rate constant for the decomposition of radium and what is its half-life?

7. In what period of time will a sample of P^{32}, which has a half-life of 14·3 days, lose 95 per cent of its reactivity?

8. Derive an expression for the average lifetime of a radioactive atom in terms of the half-life of the material.

9. The disappearance of an unsaturated hydrocarbon in the gas phase at 300°C and an initial pressure of 500 mm had a half-life of 15 min. When the initial pressure was 50 mm the half-life was 150 min. What is the order of the reaction and what is its rate constant?

10. How many half-lives must elapse before a first-order reaction is 99·5 per cent complete?

11. In the first-order reaction system A$\rightarrow$B it is possible to analyse the mixture for the amount of A it contains to an accuracy of ± 1 per cent of the initial concentration of A. With what accuracy can the rate constant be determined if the analysis is performed after the reaction has proceeded to (i) 10 per cent, (ii) 50 per cent, (iii) 90 per cent of completion? Assume that there is no error in the measurement of time.

12. The half-life for the second-order dimerization of trifluorochloroethylene is 1250 sec at 395°C when the initial pressure is 400 mm. What initial pressure will give the same half-life at 440°C if the activation energy for the reaction is 26·3 kcal mole^{-1}?

13. At 400°C the decomposition of NO_2 into NO and O_2 proceeds to completion with a second-order rate constant of 3 mole^{-1} l. sec^{-1}. If 300 mm of NO_2 are introduced into a flask at 400°C, calculate how long it will take for the pressure to reach 350 mm.

14. Chemical reactions are often "stopped" by sudden dilution of the reaction mixture. What would be the effect of a tenfold dilution on the rate of a reaction which is (*a*) first order with respect to one reactant, (*b*) second order with respect to one reactant, (*c*) third order with respect to one reactant?

CHAPTER 2

15. The following rate constants were found for the isomerization of 1,2-dimethylcyclobutene to 2,3-dimethylbuta-1,3-diene:

Temp. (°C)	149·7	168·3	182·2	197·0
$10^4 k$ (sec^{-1})	0·162	0·991	3·50	12·25

Find the Arrhenius equation by graphical means.

16. Find the activation energy of the following reactions:

(*a*) The rate constant is doubled by a change of temperature from 20° to 30°C.

(*b*) The rate constant is doubled by a change of temperature from 250° to 260°C.

(*c*) The rate constant is doubled by a change in temperature from 1000° to 1010°C.

17. Two reactions obey the Arrhenius equation

$$k_1 = 10^{14} \exp(-55{,}000/RT)$$

and

$$k_2 = 10^{15} \exp(-58{,}000/RT)$$

Find the temperature at which their rate constants are equal.

18. The rate constant for the decomposition of di-t-butyl peroxide is 6×10^{-5} sec^{-1} at 140°C. The activation energy of the reaction is 39·1 kcal mole^{-1}. What is the *A* factor?

19. The rate constant for the dimerization of trifluorochloroethylene is 65×10^{-7} mm^{-1} sec^{-1} at 440°C the activation energy is 26·3 kcal mole^{-1}. What is the A factor?

20. The A factor for the thermal decomposition of toluene

$$C_6H_5CH_3 = C_6H_5CH_2 + H$$

can be taken as 10^{13} sec^{-1}. What is the rate constant of the unimolecular decomposition at 1000°C if the activation energy is equal to the strength of the C—H bond at 84 kcal mole^{-1}?

21. If the rate of a reaction doubles between 20° and 30°C, by what factor would the rate increase for a change from 120° to 130°C? What is the activation energy?

22. The Arrhenius equation for the decomposition of t-butyl alcohol has been found to be

$$\log k\,(\text{sec}^{-1}) = 14{\cdot}7 - (65{,}500/2{\cdot}3\boldsymbol{R}T)$$

from measurements between 450° and 520°C. This experiment is repeated using a temperature measuring device that reads 10° too low over the entire temperature range (for example 450° is read as 440°). What is the new Arrhenius equation? This problem exemplifies a difficulty that may well occur in practice where the precise calibration of a temperature scale is difficult.

23. The results of a series of measurements of a rate constant over a range of temperature are expressed by the equation

$$k = (C/T)\exp-(G/\boldsymbol{R}T)$$

How are C and G related to the constants A and E of the Arrhenius equation?

24. The results of a series of measurements of a rate constant over a range of temperature are expressed by the equation

$$k = DT^{-2}\exp-(H/\boldsymbol{R}T)$$

How are D and H related to the constants of A and E of the Arrhenius equation?

25. The bond energy in a bromine molecule is 46·1 kcal mole^{-1}. What is the longest wavelength of light capable of dissociating this molecule?

26. In the photolysis of acetone in a 200 ml reaction vessel at 200°C it is found that 1 ml N.T.P. of carbon monoxide is released in 20 min. If the quantum yield for the formation of carbon monoxide is 1, what is the number of quanta absorbed per ml per sec?

27. In the photolysis of acetaldehyde at 200°C with light of 2537 Å, 1000 erg sec^{-1} of light energy is absorbed. After 20 min 0·5 ml N.T.P. of carbon monoxide is produced. What is the quantum yield for the production of carbon monoxide?

28. The reaction of hydrogen with chlorine in a 100 ml vessel was initiated by light of 4000 Å. The chlorine absorbed 15 erg of energy per second. In 30 sec the pressure of chlorine at 20°C decreased from 200 mm to 160 mm. What was the quantum yield?

29. Calculate the theoretical yield of carbohydrate $(CH_2O)_n$ formed on a hectare of land in a 50-day growing season. Assume that the sun's radiation is 10^{-2} cal cm^{-2} sec^{-1} for 8 hours a day and that the whole area is covered with green leaves that absorb all the radiation of average wavelength 6000 Å falling on them. Assume that 10 quanta of light form 1 H_2CO unit from 1 molecule of carbon dioxide and 1 molecule of water.

30. Electricity costs 1*d.* per kWh. A mercury vapour lamp converts electricity into light with 5 per cent efficiency and 20 per cent of this light of mean wavelength 2500 Å is photochemically effective. What will be the cost of the electricity needed to produce 1 kg of a compound of molecular weight 100 by a reaction that has a quantum yield of 1?

Question	Answer
1	$3{\cdot}18 \times 10^{-4}$ sec^{-1}.
2	1284 sec.
3	$1{\cdot}25 \times 10^{-4}$ sec^{-1}.
4	$1{\cdot}14 \times 10^{-3}$ sec^{-1}.
5	79·2 %.
6	$4{\cdot}26 \times 10^{-4}$ years^{-1}; 1620 years.
7	62·1 days.
9	2nd; $1{\cdot}33 \times 10^{-4}$ mm^{-1} min^{-1}.
10	7·67 half-lives.
11	11 %; 3 %; 4·5 %.
12	115 mm.
13	$4{\cdot}08 \times 10^{3}$ sec.
15	$\log k$ (sec^{-1}) $= 13{\cdot}2 - (36{,}300/2{\cdot}3\boldsymbol{RT})$.
16	12·1; 29·8; 224 kcal mole^{-1}.
17	383°C.
18	$10^{16{\cdot}48}$ sec^{-1}.
19	755 mm^{-1} sec^{-1}.
20	$3{\cdot}71 \times 10^{-2}$ sec^{-1}.
21	1·48; 12·1 kcal mole^{-1}.
22	$\log k$ (sec^{-1}) $= 14{\cdot}45 - (63{,}800/2{\cdot}3\boldsymbol{RT})$.
25	8991 Å.
26	$1{\cdot}12 \times 10^{14}$ quanta cm^{-3} sec^{-1}.
27	70.
28	$1{\cdot}46 \times 10^{6}$.
29	3×10^{5} kg.
30	133*d.*

BOOK II

Surface and Colloid Chemistry

BY G. D. PARFITT

CONTENTS

CHAPTER 1

INTRODUCTION TO SURFACE CHEMISTRY AND THE COLLOIDAL STATE

ALTHOUGH materials which are now classed as colloids have been known for many thousands of years, scientific study of colloidal systems did not begin until the early part of the nineteenth century. Thomas Graham introduced the term *colloid* in 1861 during a course of investigations on diffusion. He showed that substances which are readily obtained in the crystalline form such as sugar, common salt, etc., diffuse through water and certain membranes much more rapidly than substances like gelatin, albumen, etc., which are not easily crystallized. To substances which exhibit little or no tendency to diffuse he gave the name *colloids*, from the Greek *kolla* meaning glue, while those which diffuse rapidly were termed *crystalloids*. Later work has shown that this distinction between crystalloids and colloids cannot be maintained in its original sense. Many colloids can be crystallized and almost all crystalloids can be prepared in the colloidal state. However, the difference in diffusion rates is an important fact. For approximately spherical particles in solution it can be shown that the rate of diffusion is inversely proportional to the radius. Therefore, Graham's experiments indicate that the *particles* in a colloidal solution must be larger than those in a true solution, a fact which was confirmed by numerous experiments early in this century.

In 1907 Wolfgang Ostwald introduced the term *disperse system* when referring to substances that are more or less uniformly distributed as the disperse phase in a continuous dispersion medium. These were classified on the basis of the particle size of the disperse phase, the colloidal zone being set between two arbitrarily chosen limits as given in Table 1.

TABLE 1

DISPERSE SYSTEMS

Coarse dispersions	Colloidal dispersions	Molecular dispersions, solutions
$>0{\cdot}1\ \mu$ $>1\times10^{-5}$ cm >1000 Å	$0{\cdot}1\ \mu$–$1\ m\mu$ 1×10^{-5} cm–1×10^{-7} cm 1000–10 Å	$<1\ m\mu$ $<1\times10^{-7}$ cm <10 Å

1 micron $= 1\ \mu = 10^{-4}$ cm; 1 millimicron $= 1\ m\mu = 10^{-7}$ cm; 1 Å $= 10^{-8}$ cm.

It is now known that colloidal properties are not strictly confined to the particle size range proposed by Ostwald, but are exhibited by systems containing particles up to approximately 1 μ in size. There are, in fact, no sharp boundaries to the colloidal range.

Ostwald further classified colloidal systems on the basis of the three states of matter, namely solid, liquid and gas. The eight types of systems are given in Table 2, including the common name and some typical examples of each system.

TABLE 2

COLLOIDAL DISPERSE SYSTEMS (OSTWALD)

Disperse phase	Dispersion medium	Name	Examples
Liquid	Gas	Liquid aerosol	Fog, mist, cloud
Solid	Gas	Solid aerosol	Smoke, dust, fume
Gas	Liquid	Foam	Froth on beer, foam on soap solutions
Liquid	Liquid	Emulsion	Milk, mayonnaise
Solid	Liquid	Sol, colloidal solution	Gold sol, paint
Gas	Solid	Solid foam	Pumice
Liquid	Solid	Solid emulsion	Opal, pearl
Solid	Solid	Solid sol	Gold, ruby, glass, certain gems

The term colloidal solution (*sol*) now has a wider significance than that given by Ostwald and includes solutions of molecules such as polymers, polysaccharides, gelatin, proteins, rubber and cellulose, which are of colloidal size and in solution show colloidal properties. Also under certain conditions soap and detergent solutions show colloidal behaviour, the particles (*micelles*) being formed by aggregation of the single molecules.

Colloidal solutions are divided into two classes depending on the interaction between the disperse phase and the medium. If the mutual affinity is small the system is said to be *lyophobic* (disliking or fearing a liquid) but if great it is *lyophilic* (liking a liquid). The terms *hydrophobic* or *hydrophilic* are used if water is the medium. Most inorganic colloids are hydrophobic whereas hydrophilic colloids usually consist of organic substances. The two classes vary widely in properties as a result of their differing degrees of interaction between particles and solvent (solvation), one of the most important of these properties being the stability.

Lyophobic sols are essentially insoluble substances dispersed as small particles throughout a medium and having associated with them a large surface area. The enormous increase in surface area on subdividing a cube of side 1 cm into cubes of colloidal dimensions is shown in Table 3.

TABLE 3

INCREASE OF SURFACE AREA OF A CUBE WITH INCREASING SUBDIVISION

Length of side	Number of cubes	Total surface area
1 cm	1	6 cm^2
1 μ	10^{12}	60,000 cm^2
0·1 μ	10^{15}	600,000 cm^2
0·01 μ	10^{18}	6,000,000 cm^2
1 mμ	10^{21}	60,000,000 cm^2

The larger the surface area the greater is the surface energy and since, from the thermodynamic point of view, the energy would be expected to decrease, the particles should combine to reduce the surface area, that is, the system is essentially unstable. Also the particles, being so small, are in constant motion due to the irregular bombardment by the molecules of the medium (Brownian motion) which gives ample opportunity for collisions. In lyophobic systems the particle charge opposes this tendency and if it is large enough will considerably reduce the number of collisions, giving rise to stability. In lyophilic systems, although the particles may be charged, the more important factor in stability is the interaction between particles and solvent.

It is clear from Table 2 that colloids occur extensively in everyday life, and an understanding of their behaviour is of great importance in agriculture, biology, chemistry, medicine, and in many branches of industry (e.g. the photographic, dye, paint, printing, textile and detergent industries).

Surface chemistry is concerned with a study of the boundary between two bulk phases in contact, generally termed the *interface.* Considering the three states of matter it is evident that five interfaces can exist, viz. between two solids, solid and liquid, solid and gas, two liquids and liquid and gas. No interface can exist between two gases since they always mix spontaneously.

At the boundary there is no sharp transition between one bulk phase and the other but an interfacial region exists which shows properties differing from those of the two bulk regions. The transition region is extremely thin, being of the order of molecular dimensions, but it plays an important role in determining the behaviour of many colloidal systems and in such phenomena as ion exchange, chromatography, detergency and heterogeneous catalysis.

CHAPTER 2

THE LIQUID–GAS AND LIQUID–LIQUID INTERFACES

SURFACE TENSION

A molecule within the body of a pure liquid is attracted on all sides by neighbouring molecules, and the forces of attraction (van der Waals forces) decrease very rapidly as the distance between the molecules increases. Thus only the interactions between the first shell or two of neighbouring molecules are important and a molecule will experience essentially symmetrical forces when it is only a few molecular diameters from the surface of the liquid. At the surface, however, the attractive forces are only acting from below and sideways tending to pull surface molecules into the interior. Work is therefore required to take molecules from the interior to extend the surface, and for an increase in area of 1 cm^2 the work required is called the *surface free energy*. The surface contracts as much as possible to reduce the energy of the system to a minimum and it is for this reason that, in the absence of other forces, drops of liquid and bubbles of gas in a liquid become spherical. This tendency for the surface to contract means that molecules in the surface act as though there were a tightly stretched elastic skin over the surface. The surface behaves as if it were in a state of tension, called the *surface tension*, which is given the symbol γ and may be defined as the force in dynes acting at right angles to any line of 1 cm length in the surface. The surface free energy is expressed in erg cm^{-2} which is numerically equal to the surface tension in dyne cm^{-1}. For water the surface tension is 72·0 dyne cm^{-1} at 20°C and therefore 72·0 ergs of work are required to increase the area of the water surface by 1 cm^2. Either concept may be used to describe surface behaviour although the surface free energy is a more fundamental property.

It should be remembered that the surface of a liquid in equilibrium with its vapour is in a very turbulent state. Molecules of the liquid are evaporating from the surface into the vapour phase and vapour molecules are condensing on the surface. Calculations based on the kinetic theory show that the life-time of a molecule in the surface of liquid water is of the order of a microsecond. There is also traffic between molecules in the surface region and adjacent layers in the interior with a similar life-time. Thus on a molecular scale the surface of a liquid is in a state of violent agitation.

A similar tension to that at the liquid–vapour interface exists at the boundary between two immiscible or partially miscible liquids, and is termed *interfacial tension*. In fact this tension exists at all interfaces between different phases and in all cases the interfacial energy is the work required to enlarge the surface of separation between the two phases. Surface tension can be regarded as the interfacial tension at the liquid–vapour interface.

Quite a number of experimental methods for the determination of surface tension are known and in certain cases they may be used to evaluate interfacial tensions at the liquid–liquid interface. A simple method, though not precise, is based on the fact that a liquid rises in a clean capillary tube if it wets the walls. Following the immersion of the clean capillary the liquid wets the inside walls thus increasing its surface area. To reduce the surface energy to a minimum the liquid immediately rises in the capillary until the surface tension is just balanced by the weight of the liquid of height h (Fig. 1). It can readily be shown that

$$\gamma = rh\rho g/2\cos\theta$$

where r is the internal radius of the capillary, ρ is the density of the liquid, and g is the acceleration due to gravity (981 cm sec^{-2}). θ is the angle of contact between the surface of the meniscus and the walls of the capillary and is usually assumed to be zero for clean glass. Although simple in principle this method does not give an exact value of the surface tension because of the uncertainty of the contact angle.

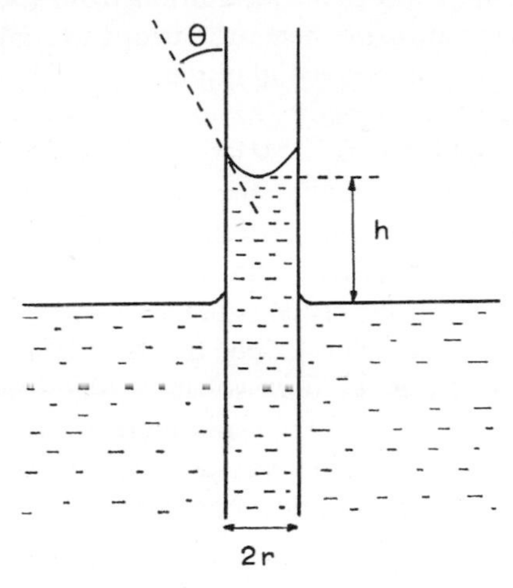

FIG. 1. Rise of liquid in a capillary tube.

For accurate measurement of surface and interfacial tension the drop-weight (or drop-volume) method is commonly used. A drop formed at the end of a fine capillary will detach itself when its weight just exceeds the upward force due to surface tension. For a drop of weight W

$$W = 2\pi r\gamma \quad \text{(approximately)}$$

r being the external radius of the capillary tip if completely wetted by the liquid. In the drop-weight method sufficient drops are collected until the weight per drop can be determined accurately. In practice the actual weight obtained is less than W because the drop breaks off at a point below the tip leaving some of the liquid behind. To allow for this a correction factor which is a function of V/r^3 is applied (V is the volume of the drop) and this can be found from tables. An alternative method is to use a micrometer syringe which measures accurately ($\pm 0{\cdot}0001\,cm^3$) the drop-volume. The drops are formed above the surface of a reservoir of the same liquid so that they are in contact with essentially saturated vapour to prevent evaporation from the surface. For interfacial tension measurements the drops are allowed to form beneath the surface of the second liquid.

ADSORPTION AT SOLUTION SURFACES

It is a well-known fact that the surface tension of solutions differs from that of the pure solvent. In 1878 J. Willard Gibbs showed on thermodynamic grounds that for a lowering of surface tension the solute concentration is higher at the surface than that in the bulk solution, and conversely, is lower if the surface tension is increased above that of pure solvent. Since the surface free energy of the system tends to a minimum then the component with the lower surface tension will be concentrated in the surface but thermal agitation prevents complete replacement of one component by the other. This surface phenomenon is generally called *adsorption* and occurs at both the liquid–vapour and liquid–liquid interfaces.

For a decrease of surface tension positive adsorption of the solute takes place and such substances are termed surface active. Many organic substances such as alcohols, fatty acids, soaps and detergents are surface active in aqueous solution and only small amounts are necessary to considerably increase their concentration in the thin surface region and hence lower the surface tension. For example, a 0·005 M aqueous solution of sodium dodecyl sulphate, a typical detergent, has a surface tension of approximately 45 dyne cm^{-1} at 25°C while that of pure water is 72·8 dyne cm^{-1}. Negative adsorption, in which case the surface tension is increased, is shown by solutions of the soluble inorganic salts, the surface concentration being lower than that in the bulk. Large amounts of the solute are required to give an appreciable effect; for instance, the surface tension of a 5 M aqueous sodium chloride solution is only about 80 dyne cm^{-1}.

Provided that the adsorption is reversible (the molecules desorb from the surface if the area is decreased) the amount of solute at the surface in excess of that in the bulk of a dilute solution is related to the surface tension lowering by the Gibbs adsorption equation

$$\Gamma = -\frac{c}{\boldsymbol{R}T} \cdot \frac{\mathrm{d}\gamma}{\mathrm{d}c}$$

where c is the concentration of the solution, T the absolute temperature, $\boldsymbol{R}$ the gas constant, γ the surface tension and Γ is the surface excess, i.e. the number of moles of solute per square centimetre at the surface in excess of that in the bulk of the solution. For dilute solutions where the surface concentration is very much greater than the bulk, Γ is the amount of solute adsorbed per square centimetre in the surface region. From the equation it follows that Γ is positive if $\mathrm{d}\gamma/\mathrm{d}c$ is negative, that is the surface tension decreases with increasing solute concentration, whereas for positive values of $\mathrm{d}\gamma/\mathrm{d}c$ negative adsorption takes place.

Thus from surface tension data the surface excess may be calculated at any concentration and from this the area A occupied per molecule in the surface is determined as in the equation

$$A = 10^{16}/N_0\Gamma \quad (\text{Å}^2\,\text{molecule}^{-1})$$

where N_0 is Avogadro's number.

There is now considerable evidence supporting the Gibbs adsorption equation. In 1932 McBain developed an ingenious method for skimming off a surface layer (0·05–0·1 mm thick) of a solution by a microtome blade travelling at high speed slightly below the surface, and this was analysed and compared with the bulk concentration. The results for both positively and negatively adsorbed solutes showed good agreement with those calculated from surface tension data using the Gibbs equation. The surface concentration has also been measured using solutes labelled with a radioisotope which emits weak beta radiation, such as carbon-14 or sulphur-35. Since the radiation is absorbed by the solution, the radioactivity measured by a detector close to the surface corresponds to that of the surface region plus only a thin layer of bulk solution. Again agreement has been found for various surface active agents with the surface excess calculated by means of the Gibbs equation.

In 1917 Langmuir suggested that the transition region between a solution and its vapour is only one molecule thick (monomolecular). This view is supported by the fact that for the surface active organic substances in aqueous solution values of Γ and A tend towards a limiting value as the concentration increases. These substances contain a hydrocarbon portion (hydrophobic) and a polar group (hydrophilic). The hydrophilic group will always tend to be in the water but the hydrophobic portion will be directed towards the vapour phase. At high concentrations the molecules are orientated vertically in a tightly packed monolayer and the observed area per molecule corresponds closely to its cross-sectional area.

SPREADING OF ONE LIQUID ON ANOTHER

If a drop of an organic liquid O is placed on the surface of water W its initial behaviour is governed by three interfacial tensions as illustrated in Fig. 2. $\gamma_{O/A}$ is the surface tension of the

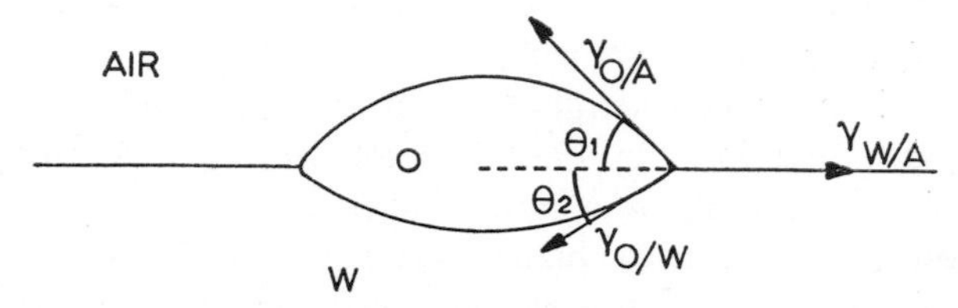

FIG. 2. Cross-section of a drop of organic liquid O lying on the surface of water W.

organic liquid and $\gamma_{O/W}$ the interfacial tension at the organic liquid–water interface acting at angles θ_1 and θ_2 to the water surface respectively. $\gamma_{W/A}$ is the surface tension of water assumed to act exactly horizontally. These tensions are, in effect, a measure of the attractive forces that molecules experience at the various interfaces and the subsequent behaviour of the drop is determined by the relative magnitude of these tensions.

If the organic liquid maintains its form as a drop the horizontal components of the tensions may be equated to give

$$\gamma_{W/A} = \gamma_{O/A} \cos \theta_1 + \gamma_{O/W} \cos \theta_2 \tag{1}$$

Such is the case for high-boiling hydrocarbons, e.g. n-hexadecane, which will not spread over the surface of water.

Consider now an organic liquid of lower surface tension. To reach an equilibrium state such that eqn. (1) applies the angles θ_1 and θ_2 must decrease and the liquid spreads to form a flattened drop for which $\theta_1' < \theta_1$ and $\theta_2' < \theta_2$ and

$$\gamma_{W/A} = \gamma_{O/A} \cos \theta_1' + \gamma_{O/W} \cos \theta_2'$$

When $\gamma_{W/A} \geqq \gamma_{O/A} + \gamma_{O/W}$ then θ_1 and θ_2 are each 0°, and the liquid spreads over the whole surface to form a duplex film, that is a film sufficiently thick such that $\gamma_{O/W}$ and $\gamma_{O/A}$ act independently. Thus,

$$\gamma_{W/A} \geqq \gamma_{O/A} + \gamma_{O/W} \quad \text{or} \quad \gamma_{W/A} - (\gamma_{O/A} + \gamma_{O/W}) \geqq 0$$

represents the condition for spreading. The term

$$\gamma_{W/A} - (\gamma_{O/A} - \gamma_{O/W})$$

is called the *spreading coefficient* S, which is positive or zero for spreading.

S may also be related to the respective works of cohesion and adhesion for the two liquids. If a column of pure liquid O of 1 cm^2 cross-section is pulled apart, two new surfaces each of 1 cm^2 cross-section are formed. The work done in pulling the liquid apart is called the *work of cohesion* W_c, for the liquid and is given by

$$W_c = 2\gamma_{O/A} \tag{2}$$

since the surface tension is also the work required to create unit surface area. Similarly, the work required to separate 1 cm^2 of interface O/W between two immiscible or partially miscible liquids against the interfacial tension is

$$W_a = \gamma_{O/A} + \gamma_{W/A} - \gamma_{O/W} \tag{3}$$

where W_a is the *work of adhesion* between the two liquids.

Combination of eqns. (2) and (3) with the definition of S gives

$$S = \gamma_{W/A} - (\gamma_{O/A} + \gamma_{O/W}) = W_a - W_c \tag{4}$$

Thus, the condition for spreading is also determined by the difference between the works of adhesion and cohesion for the two liquids. Spreading will only occur if the liquid forming the drop adheres more strongly to the water than it coheres to itself.

Since the surface tensions in eqn. (4) are for pure liquids the value of S refers to the initial state of the system before any appreciable spreading has occurred. After spreading, when the two substances are in contact, mutual saturation occurs such that their surface tensions are now changed to $\gamma'_{W/A}$ and $\gamma'_{O/A}$. The final spreading coefficient S' is defined as

$$S' = \gamma'_{W/A} - (\gamma'_{O/A} + \gamma_{O/W})$$

and the duplex film is only stable if S' is positive. In most cases S' is negative and the result is a monomolecular layer in equilibrium with excess liquid (if any) in the form of a lens on the surface. This behaviour is well illustrated by the spreading of benzene on water. For this system

$$\begin{aligned} S &= \gamma_{W/A} - (\gamma_{O/A} + \gamma_{O/W}) \\ &= 72{\cdot}8 - (28{\cdot}9 + 35{\cdot}0) = 8{\cdot}9 \text{ dyne cm}^{-1} \end{aligned}$$

and, therefore, when a drop of benzene is placed on a water surface rapid initial spreading occurs. When mutual saturation has taken place the surface tensions are reduced and

$$S' = 62{\cdot}2 - (28{\cdot}8 + 35{\cdot}0) = -1{\cdot}6 \text{ dyne cm}^{-1}$$

The final spreading coefficient is negative and the liquid retracts to form a lens leaving a monolayer of benzene on the water surface.

INSOLUBLE MONOLAYERS

The beautiful colour effects that are exhibited when drops of oil fall on water were first observed by Benjamin Franklin as early as 1765. His experiments involved the spreading of olive oil, which is insoluble in water, over the surface of a pond at Clapham Common in London, to produce very thin oil films. More than a century later Agnes Pockels developed a technique for handling insoluble oil films on a water surface using barriers resting on the edges of a trough filled to the brim, with which the area of the surface available to the oil could be varied. She observed that the surface tension was not lowered on decreasing the area until a certain critical value was reached. This was later confirmed by Lord Rayleigh in 1899 using a similar technique, and he suggested that the critical point corresponded to the oil molecules being closely packed in a monomolecular film. Hardy pointed out in 1913 that monolayers are formed by molecules consisting of a hydrophobic and a hydrophilic part and that they are orientated at the air–water interface with the hydrophilic part buried in the water, while the remainder is pushed out. This was confirmed by Langmuir in 1917 who showed that the limiting area of a series of fatty acids of different chain-lengths was the same, corresponding to identical orientations at the surface.

The apparatus used by Langmuir (the well-known Langmuir trough) has since been considerably refined, and is used for the precise study of a wide variety of insoluble monomolecular films. The basic features are illustrated in Fig. 3(*a*). A known volume of a solution of the insoluble substance in a volatile solvent is placed on a clean surface of water in the trough using a micrometer syringe. The solvent evaporates leaving a monomolecular film of the substance on the surface confined between a glass barrier and a mica float suspended from a torsion wire. To prevent leakage of the film the glass barrier, mica float, edges of the trough and the silk threads attached between the ends of the float and the sides of the trough are coated with a thin layer of paraffin wax. The film exerts a pressure on the float which causes a small rotation of the torsion wire, indicated by a beam of light reflecting from the mirror attached to the wire. By rotation of the torsion head the float is returned to its original position and the force required (in dynes per centimetre of float) is determined from a calibration

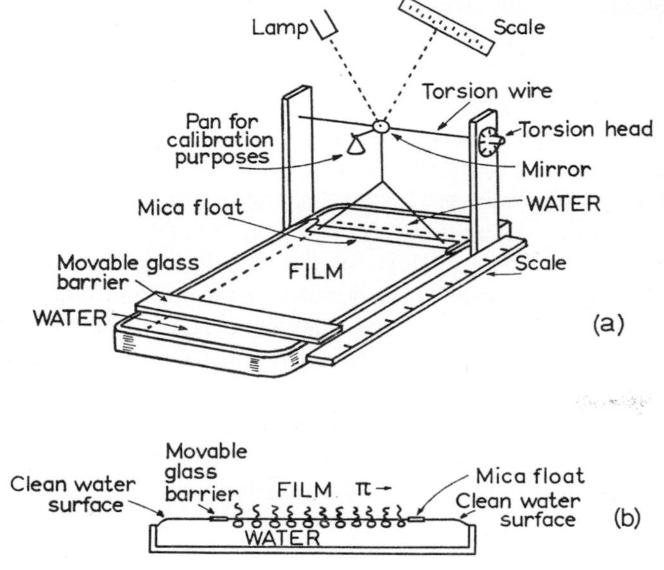

FIG. 3. The Langmuir trough for studying insoluble films. (*a*) General view of the trough. (*b*) Side-view showing the molecules (enlarged) of the film with their hydrophilic group in the water and the non-polar portion directed outwards.

graph obtained beforehand using weights in the small pan attached to the torsion wire. The force exerted on the float is due to the greater surface tension of the water beyond the float and the rotation of the torsion head is therefore a direct measure of the lowering of the surface tension by the film. This lowering is called the *surface pressure* (π) and is given by

$$\pi = \gamma_o - \gamma$$

where γ_o and γ are the surface tensions of the clean water and film-covered surfaces respectively. The molecules of the adsorbed film move freely in two dimensions and π can be regarded as being due to the bombardment of the float by these molecules (Fig. 3(*b*)). The pressure can also be regarded as an osmotic pressure, the float preventing the film from spreading over the whole surface and thus being equivalent to the membrane in osmosis. The film can be compressed by moving the glass barrier towards the float and the surface pressure measured for different film areas. Knowing the number of molecules in the film the average area A (in Å^2) occupied by one molecule can be calculated for each value of surface pressure. Plotting π versus A (sometimes called a force–area curve) gives curves which are characteristic of the film-forming substances.

The essential requirement for the formation of an insoluble monolayer on a water surface is that the molecule must be largely non-polar to give insolubility but contain a polar group conferring the necessary adhesion to the surface. The substances most commonly studied are the long-chain fatty acids (e.g. stearic acid $C_{17}H_{35}COOH$), esters (e.g. ethyl palmitate $C_{15}H_{31}COOC_2H_5$), and alcohols (e.g. cetyl alcohol $C_{16}H_{33}OH$), and also proteins and synthetic polymers. Petroleum-ether is frequently used as the volatile solvent.

Insoluble monomolecular films have length and breadth but, being only one molecule thick, can move only in two directions, that is they are two-dimensional. From their π–A relationships it was soon realized that these films could exist in various physical states some of which show close analogy with three-dimensional matter.

At very low surface pressures and large areas the molecules move about independently in the film, behaving as a two-dimensional gas. The surface pressure π and the area per molecule A correspond to the pressure P and volume V respectively of a three-dimensional gas. If the monomolecular film behaves as an ideal two-dimensional gas the equation $\pi A = \mathbf{k}T$ (or $\pi A/\mathbf{k}T = 1$) will be obeyed corresponding to the ideal gas equation $PV = \boldsymbol{R}T$

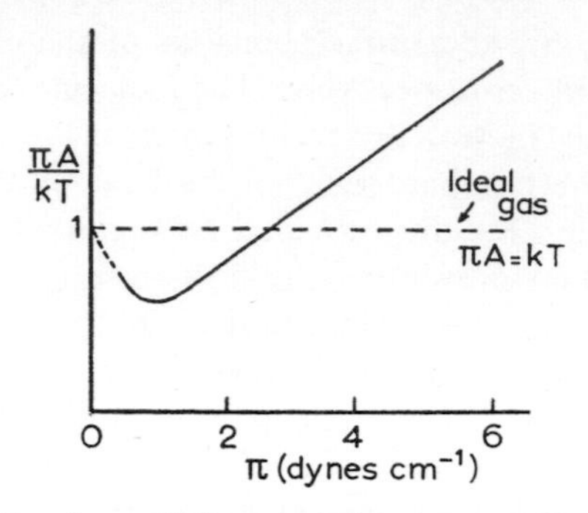

FIG. 4. $\pi A/\mathbf{k}T$ versus π for a gaseous film.

(**k**, the Boltzmann constant, is the gas constant per molecule). Figure 4 shows a $\pi A/\mathbf{k}T$ versus π curve which is obtained for a number of long-chain compounds, such as ethyl laurate $C_{11}H_{23}COOC_2H_5$, which form *gaseous* films. At very low pressure $\pi A/\mathbf{k}T$ approaches 1, but at higher pressures deviations occur which are similar to those shown by imperfect three-dimensional gases.

When the surface pressure is such that the molecules are closely packed the monolayer shows very low compressibility, that is π increases considerably for a small decrease in A. The film is termed *condensed* and corresponds to the solid state in three dimensions.

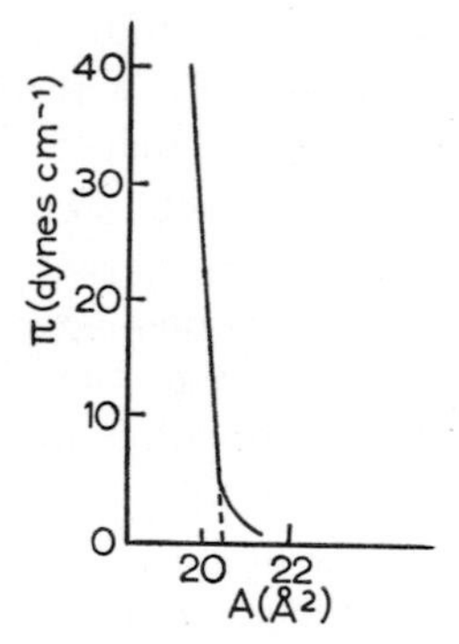

FIG. 5. Surface pressure versus area curve for a condensed film.

If talc is sprinkled over the surface of the film it is observed that whole sections of the film will move together as though it has a solid structure. The π–A curve, shown in Fig. 5, is typical of those obtained for straight chain fatty acids and alcohols. Extrapolation of the linear portion to zero pressure gives the cross-sectional area of a molecule in the film in which the molecules are close packed vertically but not deformed under pressure. This area agrees well with that obtained from X-ray measurements on

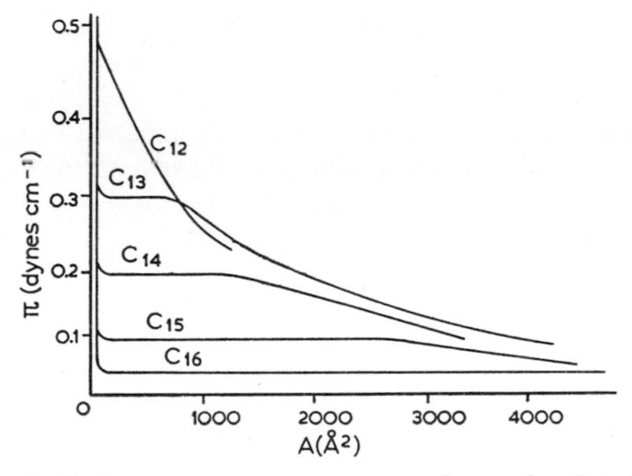

FIG. 6. Surface pressure versus area curves for a series of straight chain fatty acids.

the material in the crystalline form. For paraffin chain derivatives an area of 20–21 Å^2 is found corresponding to the cross-sectional area of the chain.

Some films exhibit transition phenomena between the gaseous and condensed states which show remarkable similarity to the condensation of vapours to liquids in three dimensions. The π–A curves for certain straight chain acids at low surface pressures show a strong resemblance to Andrews' isothermals for carbon dioxide with horizontal portions corresponding to the region of saturated vapour in the isothermals where liquid and vapour coexist (Fig. 6). In the two-dimensional film, over the horizontal portion there are regions in which the molecules interact more strongly to form a liquid-like structure separated by gaseous regions in which the molecules are moving about independently.

An important application of insoluble films is in retarding the rate of evaporation of water from reservoirs in hot climates. Cetyl alcohol spreads as a condensed film and reduces the evaporation rate by as much as 90 per cent. The alcohol is put on the water as a solid, which floats at first and then forms a film which spreads to cover the surface. The study of surface films has also led to valuable information on the structure and molecular weight of complex organic molecules such as proteins, and on the interpretation of the behaviour of emulsions. The technique has been used to study the interaction between dissolved biologically active compounds (e.g. anaesthetics) and the cell constituents, such as proteins and carbohydrates, spread as a monolayer.

CHAPTER 3

THE SOLID–LIQUID AND SOLID–GAS INTERFACES

NATURE OF THE SOLID SURFACE

Examination of the surface of a solid by electron microscopy shows that for the vast majority of solids the surface has irregularities which in a few cases may be only atomic in scale but are frequently much larger. Carefully cleaved mica is one of the rare instances in which the surface is planar. As in the case of liquids the atoms at the solid surface experience unbalanced attractive forces and therefore the solid possesses a surface tension and a surface free energy. However, atoms at the peaks of the irregularities are more energetic than those with a normal number of nearest neighbours and thus on an atomic scale, the surface tension will vary from place to place over the surface. Compared with those of a liquid the atoms at the solid surface are almost immobile and there is none of the turbulent traffic of molecules across the solid–gas interface that occurs with a liquid. When a liquid is divided into two parts the equilibrium surface tension is rapidly established, but for a similar process with a solid the atoms at the surface move very slowly to their equilibrium positions. This lack of mobility of surface atoms makes direct measurement of surface tension impossible. The surface characteristics of a solid will be determined by its past history and will be affected by such factors as heat treatment, polishing and contamination by oxide coatings, grease, etc.

Many solids also possess internal surface as a result of pores and cracks that penetrate into the bulk of the solid. Some porous solids have a much larger internal than external area, as for example the activated charcoals used frequently in organic chemistry.

THE WETTING OF SOLIDS

If a drop of liquid placed on the surface of a solid spreads over the surface the liquid is said to wet the solid. In many cases the liquid remains as a drop having a definite angle of contact θ

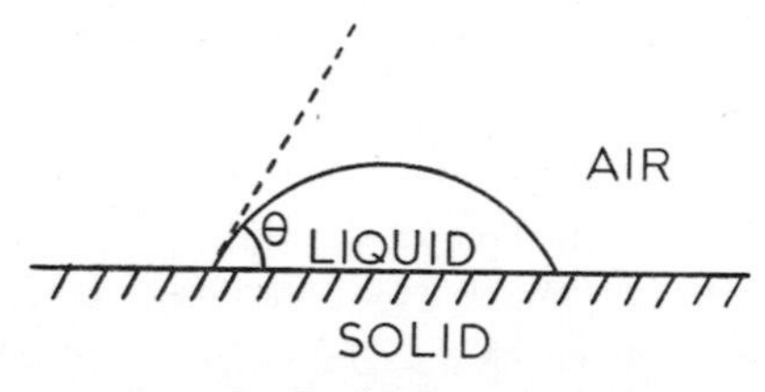

FIG. 7. Cross-section of a liquid drop resting on a solid surface with angle of contact θ.

between the solid and liquid as illustrated in Fig. 7. Wetting occurs if the mutual attraction between the atoms of the solid and the molecules of the liquid is stronger than between the liquid molecules themselves. For complete wetting the contact angle is zero; a liquid is considered not to wet the solid if the contact angle is greater than 90°. Some of the highest contact angles recorded are for mercury on glass (140°) and on steel (154°). Surface contamination may considerably modify the wetting characteristics of a solid, for example glass is normally covered with a thin film of grease and is not completely wetted by water. However, when the glass is scrupulously cleaned with chromic acid the contact angle falls to zero.

Consider a drop resting on a solid surface with a definite contact angle. Figure 8 shows a cross-section of the drop and the three forces of interfacial tension. $\gamma_{L/A}$ is the surface tension of the liquid and $\gamma_{S/L}$ and $\gamma_{S/A}$ are the interfacial tensions at the solid–liquid and solid–air interfaces respectively. Equating the horizontal components of the tensions leads to the expression

$$\gamma_{S/A} = \gamma_{S/L} + \gamma_{L/A} \cos \theta \tag{5}$$

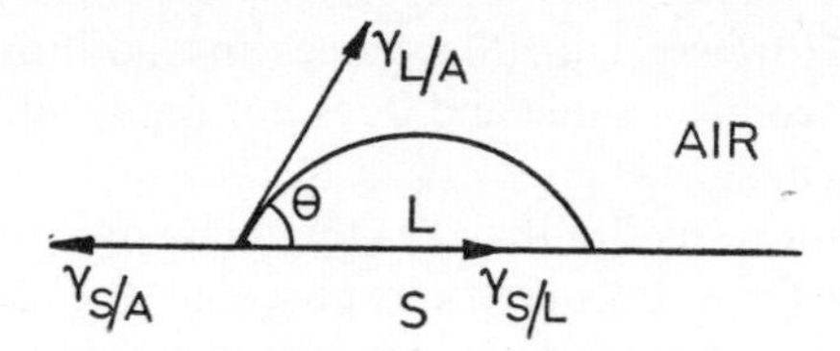

FIG. 8. Forces acting on a drop resting on a solid surface.

The work of adhesion of a solid–liquid interface $W_{S/L}$ is the work required to separate 1 cm^2 of the interface. In doing so the interface between solid and liquid is decreased by 1 cm^2 and the interfaces between solid and gas and liquid and gas are both increased by 1 cm^2, so that the work done is

$$\gamma_{S/A}+\gamma_{L/A}-\gamma_{S/L} = W_{S/L} \tag{6}$$

Combining eqns. (5) and (6) gives

$$W_{S/L} = \gamma_{L/A}(1+\cos\theta) \tag{7}$$

and thus the work of adhesion may be determined from contact angle measurements.

For the case of zero contact angle, when the liquid spreads over the surface, eqn. (7) gives $W_{S/L} = 2\gamma_{L/A}$. Now $2\gamma_{L/A}$ is the work required to pull apart a column of pure liquid 1 cm^2 in cross-section, termed the work of cohesion, $W_{L/L}$, of the liquid. Thus the contact angle depends on the relative values of $W_{S/L}$ and $W_{L/L}$ that is on the relative attraction between solid and liquid and between the liquid molecules themselves. The function $(W_{S/L} - W_{L/L})$ is called the *spreading coefficient* S of the liquid over the solid and may be expressed as

$$S = \gamma_{S/A}-(\gamma_{L/A}+\gamma_{S/L}) \tag{8}$$

If S is positive spreading will occur but if negative the liquid forms a lens on the surface. Substituting for $\gamma_{S/A}$ from eqn. (5) gives

$$S = \gamma_{L/A}(\cos\theta-1) \tag{9}$$

which suggests that S can never be positive since $\cos\theta$ cannot exceed unity. However, eqn. (8) assumes an equilibrium condition with a definite contact angle and does not apply when spreading occurs, i.e. when $\gamma_{S/A} > \gamma_{L/A} + \gamma_{S/L}$.

It is important to realize that S is the final spreading coefficient and relates to the solid surface exposed to the vapour of the liquid. In this state there is an adsorbed film of vapour on the solid surface and the value of $\gamma_{S/A}$ is usually much lower than the corresponding value for the solid exposed to vacuum. The initial spreading coefficient is therefore greater than the final and the situation can arise when the former is positive and the latter negative. In this case the liquid will at first spread over the clean surface and then retract to form a lens leaving an adsorbed film on the surface.

Measurement of Contact Angle

A number of methods have been used for contact-angle measurement but probably the most accurate values are obtained by the *tilting plate* method. The solid in the form of a smooth

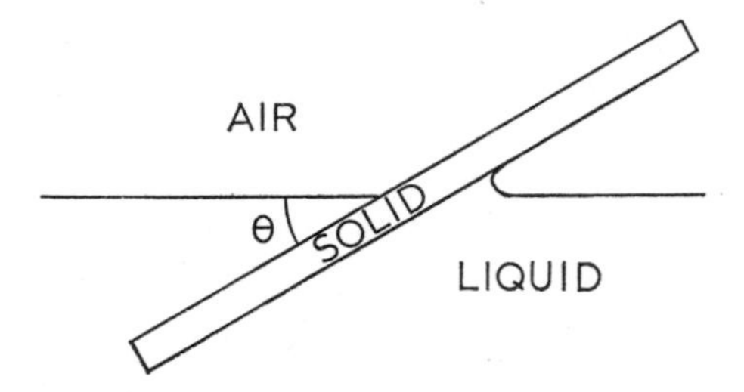

FIG. 9. A solid plate immersed in a liquid so that the meniscus on one side is horizontal.

plate dips into the liquid and its position is adjusted so that the liquid surface remains undistorted right up to the surface of the solid (Fig. 9). Alternatively the contact angle can often be measured directly by projecting on to a screen the image of a drop resting on a flat plate.

It is frequently observed that the contact angle measured on a dry surface is larger than that for the solid which had just previously been wetted by the liquid. These two values are referred to as the *advancing* and *receding* contact angles respectively and the effect is known as hysteresis of the contact angle. Several explanations have been proposed to account for this effect and no single theory appears to fit all cases. Surface roughness and contamination are probably the most important factors leading to hysteresis.

Flotation

The importance of contact angle is well illustrated by the *froth flotation* process used in the mining industry for the separation of minerals from gangue (unwanted material) and from each other. The crude ore is ground to a fine powder, suspended in water to which a substance called a collector has been added, and a stream of air bubbles blown through the suspension. The collector becomes adsorbed on the surface of the desired mineral making it more hydrophobic and thus increasing the angle of contact between mineral and water at the air–water–mineral interface. The mineral particles attach themselves to the bubbles, rise to the surface and are concentrated in the froth, while the undesirable material with low contact angle sinks to the bottom. A frothing agent is also added to stabilize the froth (see Chapter 6) so that it can be skimmed off and the mineral particles recovered. Sodium ethyl xanthate, having the structure

$$S{=}C\begin{cases} O{-}C_2H_5 \\ S{-}Na \end{cases}$$

is widely used as a collector for sulphide ores. Long-chain alcohols and pine oils are frequently used as frothing agents.

Water-repellency

Fabrics may be made water-repellent by coating the threads with a material that has a high contact angle with water, such as wax, grease and polyvalent metal soaps, which leave the exterior covered with hydrocarbon groups. Recently various silicone products have been used. The coating does not destroy the desirable air permeability of the fabric, and water would run through the gaps between the threads if sufficient pressure were applied.

The spacing of the threads is also considered as being important in water-repellency, and a high effective contact angle is possible if the threads are of small radius and quite close together. The water-repellency of the ducks' feathers is probably due to their fine structure rather than any coating that may exist on the surface. A duck will sink in a bowl of detergent solution as a result of the lowering of the effective contact angle.

Detergency

In the washing process dirt in the form of greasy material or solid particles is removed from a surface by the action of soap or detergent, and is then kept suspended in solution to prevent redeposition. In the case of the greasy dirt on a fabric the action of the detergent is to increase the contact angle and so reduce the adhesion. The grease is then removed by agitation. The detergent action is illustrated in Fig. 10 which shows how oil rolls up into globules on wool fibres by the action of detergent.

ADSORPTION FROM SOLUTION

When an aqueous solution of acetic acid is shaken with powdered charcoal some of the acid is removed by the charcoal and the concentration in solution is decreased. The process is called *adsorption* when the substance (the *adsorbate*) has been removed from solution and accumulates on the surface of the solid (the *adsorbent*). In the cases of substances that diffuse into the body of the solid the process is known as *absorption*, the solid the *absorbent*, and the substance the *absorbate*.

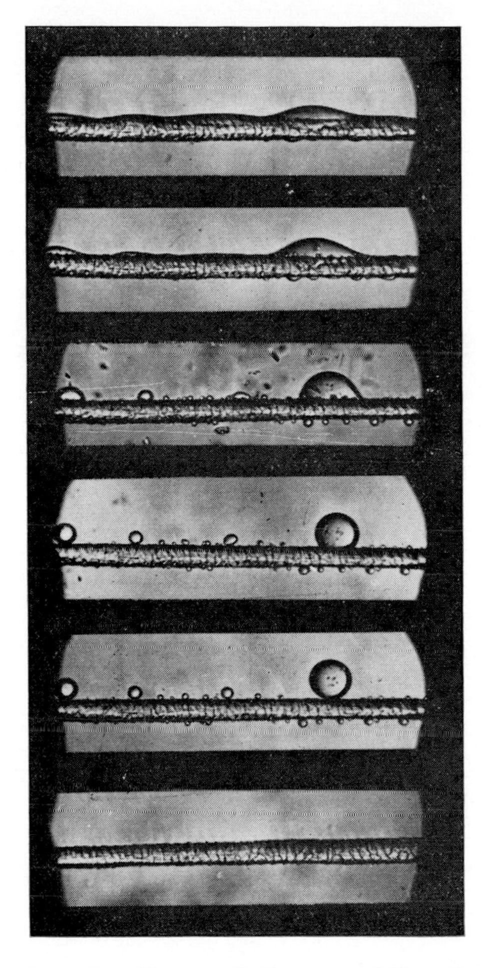

FIG. 10. Stages in rolling-up of oil on a wool fibre after addition of detergent.

Experimental investigation of adsorption from solution is comparatively simple. The adsorbent is shaken with a known volume of solution at a given temperature until there is no further change in concentration as determined by analysis of the supernatant liquid. Various techniques have been used for analysis including such physical methods as colorimetry, refractive index, surface tension, also chemical and radio-chemical methods where applicable. The apparent amount of solute adsorbed per gramme of adsorbent is then determined from the change of concentration and the experiment repeated over a wide range of solution concentrations. A plot of the amount adsorbed per gramme versus the final concentration of solution at constant temperature is called an *adsorption isotherm*. From adsorption experiments in which the change in solution concentration is determined, only an apparent adsorption isotherm is obtained since the solvent must also be adsorbed to a certain extent. The true adsorption isotherm for the solute may be calculated from the apparent isotherm but the procedure is beyond the scope of this book.

The degree of adsorption of a substance from solution at a given temperature and concentration of solution depends on the nature of adsorbent, adsorbate and solvent. For example, a polar adsorbate will be strongly adsorbed by a polar adsorbent from a non-polar solution, the polar adsorbate preferring the more polar phase. In the adsorption of an homologous series of fatty acids on polar silica gel from a non-polar solvent such as toluene the amount of acid adsorbed at any particular concentration decreases as the series is ascended. The converse is true for the adsorption of the acids on non-polar carbon from aqueous solution. Adsorption isotherms illustrating these effects are shown in Fig. 11.

Cases of negative adsorption are known in which the concentration of solute in solution is increased after shaking with the adsorbent. This occurs with dilute aqueous solutions of potassium chloride and blood charcoal powder in which case the solvent is adsorbed in preference to the electrolyte, as at the air–solution interface. However, in more concentrated solutions potassium chloride exhibits positive adsorption.

There are two main classes of adsorption depending on whether the forces between adsorbate and adsorbent are physical or chemical in character. The forces in physical adsorption are of the van der Waals type and the adsorption is reversible, that is the amount adsorbed at a given concentration and temperature is the same whether the final concentration is obtained by dilution of a stronger solution or by concentrating a weaker one. Chemical adsorption or *chemisorption* involves forces of a chemical nature

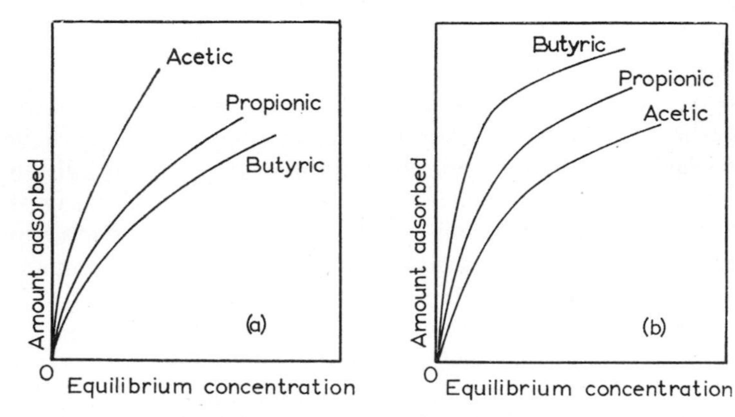

FIG. 11. Adsorption isotherms for fatty acids on (*a*) silica gel from toluene, and (*b*) carbon from aqueous solution.

similar to those concerned in chemical combination. Chemisorption is irreversible, and the adsorbate cannot be washed off the surface by treatment with the solvent. In both types the adsorption process is exothermic and the extent of adsorption decreases as the temperaure is raised. In solution physical adsorption is the more common and is observed for a variety of organic molecules (e.g. alcohols, esters, dyestuffs) in aqueous solutions on alumina, silica gel, and various forms of carbon. It has been shown that fatty acids are chemisorbed from benzene solutions on nickel and platinum catalysts.

A typical adsorption isotherm indicative of physical adsorption is shown in Fig. 12. In many practical cases it is found that the

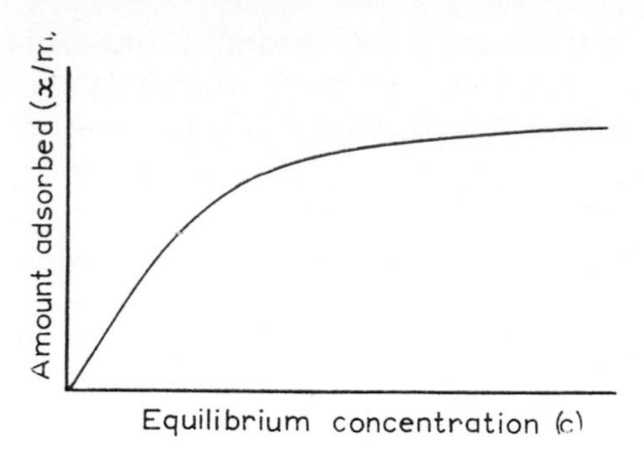

FIG. 12. Typical isotherm for adsorption of a solute at the solid–liquid interface.

adsorption obeys an equation known as the *Langmuir isotherm*. This equation was derived for adsorption of gases on solids and assumes that the adsorption is limited to a monolayer and occurs on a uniform surface. The Langmuir equation may be written as follows:

$$x/m = abc/(1+bc)$$

where x is the amount of solute adsorbed by m g of adsorbent, c is the concentration of solution at equilibrium, and a and b are constants. At high concentrations $(x/m) \rightarrow a$ when $bc \gg 1$, and a is then the amount adsorbed at saturation. At low concentrations when $bc \ll 1$ the amount of solute adsorbed is proportional to c. To test the fit of the Langmuir isotherm to experimental data the equation is put in the form

$$c/(x/m) = (1/ab)+(c/a)$$

so that a plot of $c/(x/m)$ versus c should be a straight line of slope $1/a$ and intercept $1/ab$. The surface area of the powder (usually expressed as square metres per gramme) may be obtained from the derived value of a provided that the area occupied by the adsorbed molecule on the surface is known with reasonable certainty. It must be remembered that the surface area determined by this method is the total area accessible to the solute molecules. If these are large (dyestuffs, long-chain molecules) they may not penetrate the pores and cracks, and the area obtained may be only a fraction of the true surface area of the solid.

Chromatography

A very important application of adsorption from solution, which is based on the fact that a given solid adsorbent will adsorb to varying extents the different components of a mixture in solution, is known as *chromatographic analysis* or *chromatography*. The principle was first used by the Polish botanist Tswett in 1906 for separation of leaf pigments but was not widely applied until about 1930.

The adsorbent, in powder form (alumina is commonly used), is closely packed in a column and the solution containing a mixture of solutes, dissolved in a suitable solvent, is allowed to percolate slowly through it. The compounds adsorbed in the upper layers of the column will be those with the greatest affinity for the solid surface and less strongly adsorbed components will concentrate further down the column in the order of their ease of adsorption. A better separation is achieved by then allowing some of the pure solvent to flow slowly through the column. Each component is now virtually restricted to a definite layer in the column, and these layers are readily seen when the substances are coloured. The components can be isolated by removing the column bodily and then cutting it up into sections corresponding to the various layers, from which each component is extracted separately with solvent. Alternatively the elution procedure may be used in which more solvent is passed through the column until each layer reaches the bottom and leaves the column as a solution, from which the component is isolated or successive fractions identified by means of some physical property such as conductivity or refractive index.

Chromatography is now used extensively for analytical and preparative work with both inorganic and organic compounds in the laboratory and on a small scale in industry. In many cases, particularly in biological work, filter paper has replaced the packed column. Mixtures of gases can also be separated using as solvent a carrier gas (usually nitrogen) and either a solid adsorbent or a high-boiling liquid immobilized on a solid support (for example a silicone fluid on kieselguhr).

Ion Exchange

When hard water is passed through a column of a natural zeolite the calcium ions in the water exchange with the sodium ions of the solid thus softening the water. The reaction may be represented by the equation

$$2Z-Na^{+}+Ca^{++}(\text{soln.})\rightleftharpoons Z_2Ca^{++}+2Na^{+}$$

where Z is the zeolite. Alternatively the surface of the zeolite may first be saturated with hydrogen ions (by passing dilute acid through the column) and these exchange with the cations in the water.

A number of highly porous synthetic resins are commercially available for both cation and anion exchange. One of each type is required for the demineralization of water, the first being saturated with hydrogen ions and the second with hydroxyl ions. These ions are exchanged with the ions in the impure water and combine to form water. The resins become exhausted with time but can be regenerated by passing dilute acid through the cation exchanger and dilute alkali through the anion exchanger.

There are many commercial applications of ion exchange, the production of pure water being one of the most important. In the laboratory Spedding successfully separated the rare earths on a milligram scale by passing a solution of the mixed oxides in hydrochloric acid down a column of a cation exchange resin. The hydrogen ions on the resin exchange with the rare earth cations and elution is then carried out with a citric acid solution, which forms ionic complexes with the elements with quite different adsorption behaviour.

ADSORPTION OF GASES AND VAPOURS ON SOLIDS

All solids adsorb gases to some extent but the effects are not very evident unless the adsorbent possesses a large surface area for a given mass. Porous materials such as silica gel and charcoals are very effective adsorbents.

Several methods have been used for obtaining gas adsorption data. Probably the most common method is to expose the solid, outgassed by heating under vacuum (to remove previously adsorbed gases), to a known amount of gas and to determine the pressure before and after adsorption. Knowing the volume of the apparatus not occupied by the adsorbent the amount of gas adsorbed at a given pressure can be calculated from the gas laws. Further quantities of gas are then admitted and the amount of gas adsorbed at various equilibrium pressures determined. Alternatively the amount of gas adsorbed may be determined by direct weighing. In this method the outgassed solid is contained in a small platinum bucket attached to a quartz spiral spring. The extension of the spring is measured and related directly to the increase in weight due to adsorption by means of a previous calibration using small known weights.

Two types of adsorption are recognized, namely physical adsorption and chemical adsorption (chemisorption) depending on whether the gas molecules are attached to the surface by van der Waals forces or by forces similar to those involved in chemical combination. Adsorption is an exothermic process and the heat evolved per mole of adsorbate is called the *heat of adsorption.*

Physical adsorption takes place very rapidly at all temperatures and the process is reversible, the adsorbate being removed in a chemically unchanged condition when the pressure is lowered. The relatively small heats of adsorption, about 5 kcal per mole of adsorbate or less, are of the same order as heats of vaporization. A comparison of the relative amounts of various gases adsorbed by a particular adsorbent at a given temperature and pressure shows that the extent of adsorption parallels the increase in critical temperature and boiling point of the gases. This is shown by the data in Table 4, where the volumes of different gases (reduced to N.T.P.) adsorbed by 1 g of charcoal at 25°C and 760 mm pressure are given together with the physical constants of the gases.

TABLE 4

ADSORPTION OF GASES BY CHARCOAL AT 25°C

Gas	Volume adsorbed (cm^3 at N.T.P.)	Critical temperature (°C)	Boiling point (°C)
H_2	1·8	−239	−259
N_2	10·9	−146	−210
CO	14·1	−141	−191·1
CO_2	60·1	31	−78·2
N_2O	67·0	38·8	−89·8
NH_3	136	130	−33·5

This correlation suggests that the most easily liquefiable gases are the more readily adsorbed and that the forces involved in adsorption are the same as those in condensation, that is van der Waals forces. The amount of gas adsorbed is not a function of its chemical properties.

Chemisorption is characterized by a much greater heat of adsorption than that for physical adsorption (usually) and is of the order of the heat of chemical reaction (20–100 kcal $mole^{-1}$). The rate of adsorption often increases rapidly with temperature indicative of a process involving an activation energy. At low temperatures chemisorption may be so slow that experimentally only physical adsorption is observed, but as the temperature is raised chemisorption becomes apparent when the adsorbed molecules acquire sufficient energy to interact chemically with the surface. This behaviour is observed with hydrogen on nickel powder. Chemisorption is often irreversible, the adsorbed molecules being difficult to remove by merely reducing the pressure, and when desorption does take place it may be accompanied by chemical changes as in the case of oxygen adsorbed on charcoal at room temperature which comes off as carbon monoxide and carbon dioxide with strong heating. In physical adsorption multimolecular layers of adsorbed gas molecules occur at temperatures below the critical temperature and at pressures approaching the saturation vapour pressure of the liquid, whereas chemisorption is always confined to a single atomic or molecular layer.

Physical Adsorption

The adsorption isotherm relating the amount of gas adsorbed to the equilibrium pressure at a given temperature is usually represented graphically and for vapours it is preferable to express the results in terms of the relative vapour pressure p/p_0, where p is the equilibrium pressure and p_0 the saturation vapour pressure of the liquid at the temperature concerned. Five general types of isotherms have been observed and are shown in Fig. 13 with examples of each type.

Type I is the Langmuir type and shows the adsorption reaching a limiting value thought to correspond to a monolayer. Type II (very common) and Type III (comparatively rare) show multilayer formation. Types IV and V in which a limit is reached below the saturation pressure are generally considered to indicate condensation in the capillaries and pores of the solid, since the vapour pressure above the concave surface of the liquid in a capillary is less than the saturation vapour pressure measured over a plane surface. Only Type I isotherms are found with gases above their critical temperature.

An equation for Type I isotherms was derived theoretically by Langmuir in 1918 and is called the *Langmuir adsorption isotherm.* Adsorption is assumed to be restricted to a monolayer on a uniform surface and at any given pressure there is a dynamic equilibrium set up such that the rate at which molecules strike the surface equals the rate of evaporation off the surface. Interaction

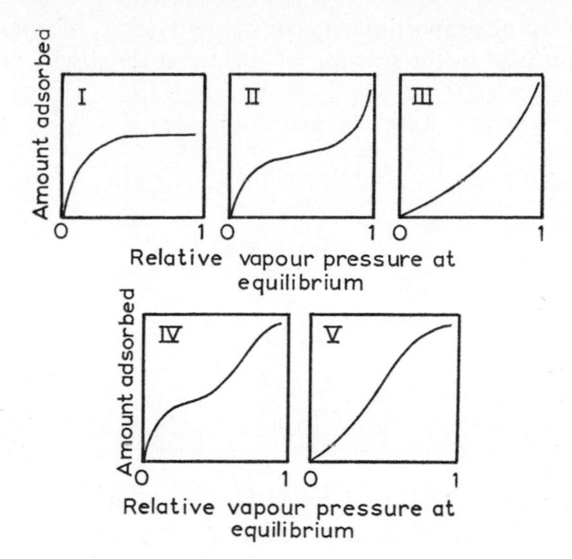

FIG. 13. The five types of adsorption isotherms. Examples:

Type I	Nitrogen on charcoal at −195°C
Type II	Nitrogen on alumina at −195°C
Type III	Bromine on silica gel at 79°C
Type IV	Benzene on ferric oxide gel at 50°C
Type V	Water vapour on charcoal at 100°C

between adsorbed molecules is assumed to be negligible and the heat of adsorption is therefore independent of coverage. The Langmuir equation may be derived as follows:

Suppose that at a given pressure of gas p, the fraction of the total surface covered with adsorbed molecules is α, then the fraction $(1-\alpha)$ is available for further adsorption. According to the kinetic theory the rate at which molecules strike that part of the surface not covered by adsorbed molecules is proportional to the pressure of the gas and to the fraction of surface available. Thus

$$\text{rate of adsorption} = k_1 p(1-\alpha)$$

where k_1 is the proportionality constant. The rate of evaporation is proportional to the fraction of surface covered with adsorbed molecules

$$\text{Rate of evaporation} = k_2 \alpha$$

When equilibrium is established these two rates are equal and therefore

$$k_1 p(1-\alpha) = k_2 \alpha$$

or

$$\alpha = \frac{k_1 p}{k_2 + k_1 p}$$

$$= \frac{bp}{1+bp}$$

where $b = k_1/k_2$.

The amount of gas x adsorbed by m g of adsorbent must be proportional to the fraction of surface covered, and hence

$$x/m = k\alpha = \frac{kbp}{1+bp}$$

or, in the linear form,

$$p/(x/m) = (1/kb) + (p/b)$$

At low pressures $bp \ll 1$ and the amount adsorbed will be directly proportional to the pressure, while at high pressures $bp \gg 1$ and the amount adsorbed is independent of pressure. This behaviour is shown by the Type I isotherm. A test of the fit of the Langmuir equation to experimental data is a plot of $p/(x/m)$ versus p, and in many cases for which a Type I isotherm is observed a good straight line is obtained, lending support to the idea that the adsorption is restricted to a monolayer.

The shapes of the Types II and III isotherms were interpreted by Brunauer, Emmett and Teller in 1938 on the basis of multi-layer adsorption at higher relative pressures. They derived an equation (now known as the B.E.T. equation) assuming that the state of dynamic equilibrium postulated by Langmuir holds for each molecular layer and that the heat of adsorption for all layers beyond the first is equal to the latent heat of condensation of the adsorbate. Isotherms of Type IV and V have been explained by extending the theory of multimolecular adsorption to include condensation in the capillaries at pressures below the saturation value.

Chemisorption and Catalysis

Chemisorption occurs when the temperature is above the critical temperature of the adsorbate or is such that the pressures involved are much lower than the saturation pressure, and under these conditions physical adsorption is usually negligible. The adsorption involves either exchange or a sharing of electrons between the adsorbate and adsorbent. The adsorbate may exist on the surface as a cation (e.g. sodium vapour on tungsten) or an anion (e.g. oxygen on nickel oxide), or may form a covalent bond (e.g. hydrogen on tungsten) or a coordinate bond with the surface. Chemisorption is always confined to a monolayer and in a number of cases is satisfactorily described by the simple Langmuir equation. This equation assumes a heat of adsorption independent of coverage and deviations have been explained on the basis of a non-uniform surface containing adsorption sites of varying degrees of interaction with the adsorbate. The most active sites with the largest heat of adsorption will be occupied first, and with further adsorption the heat decreases. Equations have been derived which take account of this behaviour.

A large number of reactions between gases proceed at a significantly increased rate in the presence of a solid which acts as a catalyst. Well-known examples in chemical industry are the Contact process for the manufacture of sulphuric acid and the Haber ammonia process. The catalytic effect of solids was attributed by scientists in the nineteenth century to adsorption of the reactants on the surface of the solid, which increases their concentration and consequently the probability of reaction would be increased. However, the fact that physical adsorption alone is not sufficient for catalysis to occur and that a given catalyst may affect the rate of one type of reaction and not another, has led to the generally accepted opinion that chemisorption is always involved in catalysis with additional physical adsorption in some cases.

CHAPTER 4

ELECTRICAL PHENOMENA AT INTERFACES

ONE of the earliest demonstrations of the existence of a potential difference associated with the interface between two phases in contact was made by the Russian scientist Reuss in 1809. The bottom of a U-tube was packed with clay, the arms filled with water and a potential applied across the clay plug by two metal electrodes immersed in the water. After a while he observed that the water level was higher in the arm containing the cathode and any suspended clay particles moved towards the anode. This would appear to indicate that on bringing the two phases into contact some positive charges had been transferred from the clay to the water, perhaps by dissociation of groups on the surface of the clay, with the result that the water now carrying an excess positive charge, moved towards the cathode. The clay particles are then negatively charged and move in the opposite direction. An alternative explanation would be that hydroxyl ions from the water are adsorbed on the clay surface which is then negatively charged with respect to the water. The movement of a liquid relative to a fixed solid under an applied field is known as *electro-osmosis* and almost invariably occurs when a potential is applied across a porous diaphragm immersed in a liquid. In the case of the solid particle moving with respect to the liquid when a field is applied, the process is called *electrophoresis*. A phenomenon which is essentially the reverse of electro-osmosis was discovered by Quinke in 1859. If the liquid is forced through a porous diaphragm a potential difference, known as the *streaming potential*, is set up between the two sides of the diaphragm. The situation in electrophoresis may also be reversed by allowing the charged particles to settle under gravity (or move in a centrifugal field), when a potential difference is set up between the top and bottom of the tube. This is termed *sedimentation potential* or the *Dorn effect*, after its discoverer. These four phenomena have in common the fact that an electrical potential difference is associated with the relative motion between a surface and a liquid, and are called *electrokinetic phenomena.*

THE ELECTRICAL DOUBLE LAYER

The nature of the distribution of charges at the solid–liquid interface giving rise to the potential difference associated with electrokinetic phenomena was first considered by Helmholtz in 1879, who introduced the concept of the electrical double layer.

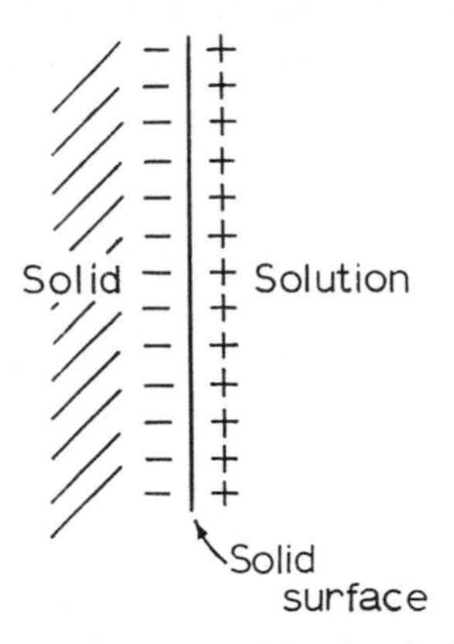

FIG. 14. The simple picture of the electrical double layer.

Two electrically charged layers of opposite sign were assumed, one fixed at the solid surface and the other close to the surface in the mobile liquid phase (Fig. 14). The charge on the surface layer is exactly balanced by an equivalent number of oppositely charged ions in the liquid layer since the double layer as a whole must be electrically neutral. Within the double layer the potential drops to zero, and the rest of the liquid phase is electrically neutral since it must contain an equivalent number of positive and negative ions. In 1904 Perrin suggested that the two layers could be considered as being equivalent to a parallel plate condenser.

This simple picture of the double layer, however, does not take into account the thermal motion of ions in solution, which would prevent the formation of such a compact arrangement. Consider a plane surface uniformly covered with negative charges in contact with an aqueous solution containing positive and negative ions. The positive ions will be attracted by the surface charge and tend to concentrate near the surface, thus reducing the electric field intensity due to the surface at points further out in solution. Hence, the concentration of positive ions will decrease, rapidly near the surface and more slowly further away. Negative ions will be repelled by the surface and thus their concentration will be reduced particularly in the vicinity of the surface. The net result is that close to the charged surface there will be an excess of positive over negative ions, the difference decreasing with distance until the point is reached in bulk solution where there is an equivalent number of positive and negative ions to maintain electroneutrality. The effect of thermal agitation is, therefore, to create a *diffuse* double layer, with the total net charge in the solution part balanced by the equal and opposite charge on the surface. The potential ψ within the diffuse double layer decreases with distance from its value at the surface (ψ_0), rapidly at first then more and more slowly until it is zero in bulk solution. This picture of the diffuse double layer due mainly to Gouy and Chapman during the years 1910–17, is illustrated in Fig. 15.

The effect of varying the parameters of the system on the ionic distribution and the decay of the potential will now be considered.

(a) *Effect of surface potential* ψ_0. For a given concentration of ions of given valency the shape of the potential decay curve is fixed. The origin of the curve is determined by the value of ψ_0. At high surface potentials of the order of 100–150 mV, the rate of decay of ψ over a particular region close to the surface is much greater than when ψ_0 is low, say 25 mV. The reason for this is that the negative ions are almost completely expelled from the region close to the surface when the potential is high and the concentration of positive ions in this region is correspondingly increased. The structure of the diffuse double layer is then determined essentially by the ion of opposite sign to the charge on the surface. This is generally the situation in colloidal systems.

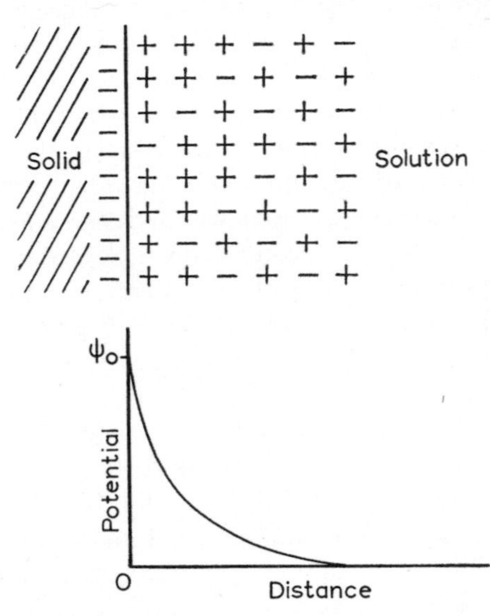

FIG. 15. Schematic representation of the diffuse double layer and the decay of potential with distance from the plane surface.

(b) *Effect of ionic concentration and valency.* An increase in concentration of the ions in solution will, at constant surface potential, mean a proportional increase in their concentration near the surface and a faster rate of decay of potential. The distance to which the diffuse part of the double layer extends from the surface is also reduced, that is the double layer is compressed. Increase in valency will have the same effect by increased attraction with the positive ions and increased repulsion with the negative ions, although at high surface potential the latter is unimportant. Potential decay curves illustrating these effects are shown in Fig. 16.

To simplify the mathematical treatment of electrokinetic phenomena it is useful to consider all the ions in the diffuse layer to be located in a single plane at a distance d from the surface. The diffuse double layer is now equivalent to a parallel plate

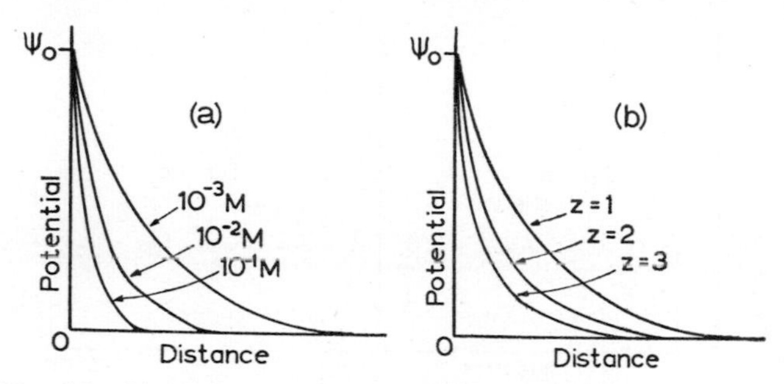

FIG. 16. (*a*) Effect of concentration of univalent ions upon the decay of potential. (*b*) Effect of valency z of ions at constant concentration (10^{-3} M) upon the decay of potential.

condenser similar to that proposed by Perrin, and called the *Helmholtz double layer* in honour of the theoretical contributions made by Helmholtz on electrokinetic phenomena. The distance d is called the *thickness of the double layer* and is related to the concentration and valency of the ions in solution by

$$d = \sqrt{(\varepsilon \mathbf{k} T / 4\pi e^2 \sum nz^2)}$$

where ε is the dielectric constant of the medium, $\mathbf{k}$ the Boltzmann constant, T the absolute temperature, e the electronic charge, n the number of ions per cm^3 and z the valency of the ions. The sign $\sum$ means that the product nz^2 is summed over all the different ionic species present in solution.

In the case of an aqueous solution of univalent ions at 25°C approximate values of d are 10 Å in a 0·1 M solution, 100 Å in a 10^{-3} M solution and 1000 Å in a 10^{-5} M solution.

The only interaction so far considered between the charged surface and the ions in solution is due to electrostatic forces, and the thermal agitation is sufficient to overcome these forces and keep the ions in solution. In certain cases, however, the ions may be attached to the surface by particularly strong electrostatic forces as with highly charged ions (Th^{++++}), or when van der Waals

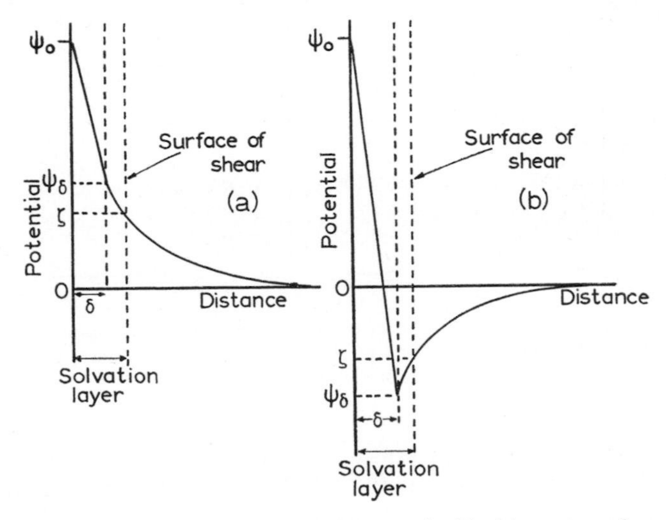

FIG. 17. Potential decay curves for a double layer at a plane surface. (*a*) $\psi_0 > \psi_\delta > \zeta$ part of the diffuse layer is located within the solvation layer. (*b*) $\psi_0 > \psi_\delta > \zeta$ but with reversal of sign of ψ_δ and ζ due to strong adsorption in the Stern layer.

forces reinforce the electrostatic forces (large organic ions). Stern in 1924 suggested that the region near the surface should be divided into two parts, the first a compact layer adjacent to the surface containing any adsorbed ions and the second consisting of the diffuse Gouy layer. The compact part of thickness δ is called the Stern layer and the potential at the boundary between the Stern and Gouy layers is ψ_δ. In this theory Stern also allowed for the dimensions of the ions which had hitherto been considered as point charges.

In addition to any ions that may be adsorbed the surface will be solvated by a layer of solvent molecules, probably of the order of one molecule thick, and there as a result of the attractive influence of the surface charge on the water dipoles. The boundary between the rigidly held solvation layer and the solution is the *surface of shear* associated with electrokinetic phenomena, and the potential at this boundary is known as the *zeta potential* ζ. In most cases ζ will be less than ψ_δ (unless large ions are adsorbed) and part of the diffuse layer will be located within the solvation layer. The modern picture of the double layer is illustrated in Fig. 17 which shows two situations that may arise.

The picture of the double layer at a curved surface, such as that associated with a spherical colloidal particle, is similar to that for a plane surface although there are quantitative differences.

ELECTROKINETIC PHENOMENA

Electrokinetic phenomena involve the relative motion between a charged surface and the diffuse double layer. This motion may be caused either by application of an electric field (electrophoresis and electro-osmosis) or by mechanical forces (streaming and sedimentation potentials). Since the relative motion occurs at the surface of shear the results of electrokinetic measurements may only be interpreted in terms of the zeta potential at this shear surface and give no direct information about ψ_0 and ψ_δ.

Electro-osmosis and Streaming Potential

In electro-osmosis the charged surface is fixed and the mobile diffuse layer moves under an applied field carrying solvent with it. The double layer may form as a result of ionization of the material of the surface or through preferential adsorption of ions from solution. A simple apparatus that may be used for the measurement of electro-osmotic flow is shown in Fig. 18. The porous plug separating the two reservoirs containing the liquid phase must have pores of radius greater than the thickness of the double layer. On application of the electric field across the plug the ions in that part of the diffuse layer outside the surface of shear move to the appropriate electrode and since they are solvated there is a net flow of liquid towards the electrode of opposite sign to that

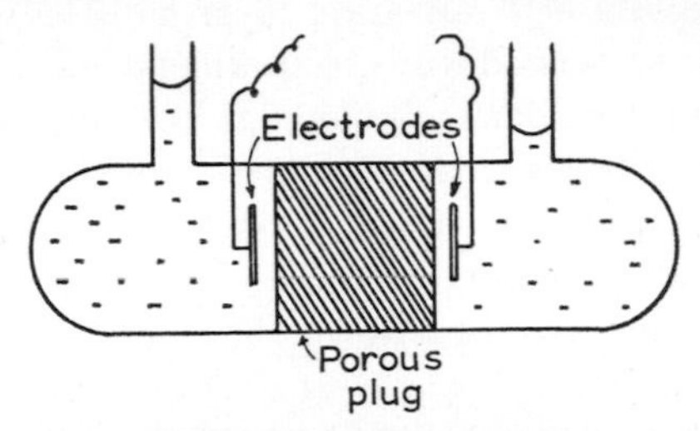

FIG. 18. Apparatus for the measurement of electro-osmosis.

of the charge of the diffuse layer. A difference in levels develops in the vertical tubes and the hydrostatic head causes a counterflow of liquid through the centre of the pores. This is illustrated in Fig. 19 which shows an enlarged view of one pore. A steady state

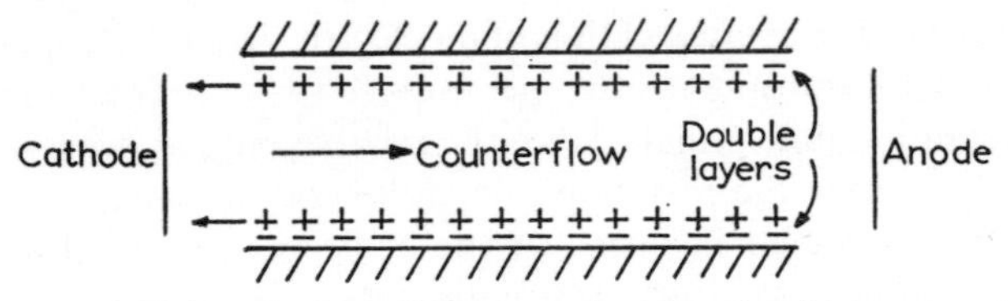

FIG. 19. Enlarged view of one pore showing the electro-osmotic flow.

electro-osmotic pressure is finally established when the counterflow just balances the flow at the pore walls. Using the concept of the Helmholtz double layer the pressure P (obtained from the difference in levels in the vertical tubes) can be shown to be related to the zeta potential by the equation

$$P = 2\zeta\varepsilon E/\pi R^2$$

where E is the potential applied across the plug containing pores of radius R, and ε is the dielectric constant of the medium. For water at 25°C an applied potential of 1 V will give a pressure of 1 cm of mercury in pores of approximately 10,000 Å in radius.

Electro-osmosis does not have wide industrial application although the technique has been used to remove liquid from porous materials such as peat and clay.

By forcing the liquid through the porous plug the situation in electro-osmosis may be reversed and a potential, the streaming potential E_s, is set up across the electrodes. It can be shown that

$$E_s = \zeta P\varepsilon / 4\pi\eta\kappa$$

where P is the pressure forcing the liquid through the plug, η is the viscosity and κ the specific conductivity of the liquid. Considerable streaming potentials can arise when liquids of low conductivity are used, for example when petroleum is pumped through metal pipes. The risk of explosion is reduced by addition of agents that ionize to a small extent in the hydrocarbon and increase the conductivity.

Electrophoresis and Sedimentation Potential

Electrophoresis is the converse of electro-osmosis. In electrophoresis a charged particle moves with respect to an immobile liquid under the action of an electric field. This effect can easily be observed using a coloured colloidal dispersion such as manganese dioxide, in the apparatus shown in Fig. 20. The U-tube is first partly filled with an electrolyte solution of similar conductivity to that of the colloidal dispersion, and the colloid is then run in slowly so that sharp boundaries are formed in the two arms of the tube. On application of a potential across the electrodes the boundary in one arm slowly rises and that in the other falls according to the sign of the charge at the surface of shear. The zeta potential in volts may be determined from the velocity of migration of the boundary v from the following equation

$$v = (\zeta \varepsilon X / 4\pi\eta) \times (1/300)^2$$

where X is the applied potential gradient (in volts per centimetre). For an applied potential of E V between electrodes L cm apart, $X = E/L$. The factor $(1/300)^2$ allows practical units (volts) to be used in place of electrostatic units (1 V = 1/300 e.s.u.). This equation is accurate for low values of ζ and large particles, but in other cases corrections must be applied which allow for the fact that the net charge in the mobile part of the double layer is

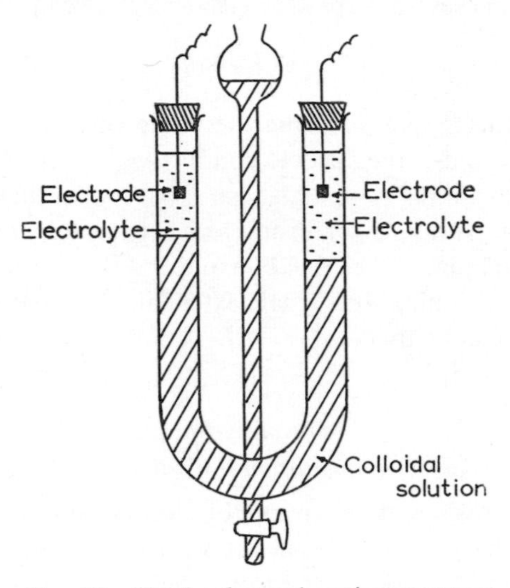

FIG. 20. U-tube electrophoresis apparatus.

opposite to that at the surface of shear, and moves in a direction opposite to that of the particle. These ions carry solvent with them so that there is a flow of medium opposing the motion of the particle and thus reducing its velocity.

The technique of microelectrophoresis may be used for determining the electrophoretic mobility of particles which are visible using a microscope. A useful type of microelectrophoresis cell consists of a single horizontal tube of rectangular cross-section the distance between the top and bottom inside the cell being about 1 mm. The velocity of the particles is determined by timing their movement across a calibrated graticule in the eyepiece of the microscope. It is necessary to allow for the electro-osmotic flow of liquid which occurs near the glass walls and for the counterflow in the centre of the tube. There are two levels within the cell at which these are just balanced and only by measuring the particle velocity at these levels can the true electrophoretic mobility be determined.

Electrophoretic measurements have provided important information on the stability of colloidal dispersions and the technique has wide application in the fractionation and analysis of biological molecules.

The settling of charged particles in a liquid under gravity, or if small when centrifuged, gives rise to the sedimentation potential, which may be measured by two electrodes placed near the top and bottom of the vessel. This is the converse of streaming potential and is the least important of the electrokinetic phenomena. Significant sedimentation potentials may arise when water droplets in petroleum, formed during the pumping operation, settle under gravity in large tanks after filling. The danger of explosion is removed if substances are added to increase the conductivity of the petroleum.

CHAPTER 5

COLLOIDAL SOLUTIONS

COLLOID science is concerned with systems that contain one component in a finely divided state dispersed in the other as a continuous medium. The term *colloidal solution* or *sol* refers to systems in which the dispersion medium is a liquid, and the dispersed particles are either solid or are large molecules whose dimensions are in the colloid range (10–10,000 Å approx.). Dispersions of a gas in a liquid (foam) and a liquid in a liquid (emulsion) will be discussed in Chapter 6.

Colloidal solutions may be roughly divided into two main types known as *lyophobic* (liquid-hating) and *lyophilic* (liquid-liking). If water is the dispersion medium the terms *hydrophobic* and *hydrophilic* are sometimes used. The classification is based on the affinity of the disperse phase for the medium which in the case of the lyophobic sols is very small, as in dispersions of various metals and silver halides in water. In lyophilic sols, on the other hand, there is a strong affinity between the disperse phase and the medium. Examples of lyophilic sols are solutions of polymers, proteins and soaps. The two extreme types show widely differing properties, the most important of which is the stability. Lyophobic sols are sensitive to the addition of small quantities of electrolyte, which allow the particles to stick together (flocculate) and separate from the dispersion medium, whereas lyophilic sols are unaffected by electrolytes at low concentration. In the former type the electrical repulsion between particles is reduced by the electrolyte, whereas in the lyophilic systems the solvation of the particles is the important stability factor. Although the classification is not absolute since there are sols that show intermediate properties, it is convenient to divide colloidal solutions into the two types and to consider their behaviour separately.

LYOPHOBIC SOLS

The following account deals with the preparation and properties of dilute dispersions of solid colloidal particles in aqueous media, since most of our present knowledge of lyophobic sols comes from fundamental research on such systems. The particles are assumed to be charged, the origin of which charges will be discussed, and stability arises from the mutual repulsion between the particles due to the charge.

Optical Properties

Lyophobic sols may be coloured or colourless. A gold sol is cherry-red if the particles are small, the colour being due to absorption of green light. Colourless colloids absorb little or none of the wavelengths of the visible spectrum (4000–7000 Å). However, all colloidal solutions exhibit the phenomenon known as *scattering*, which in sufficient concentration makes them appear turbid in normal daylight. If the particles are small the solution frequently looks clear but the scattered light is readily observed by passing an intense beam of white light through the colloidal solution and viewing from the side against a dark background. The appearance of the beam of light in the solution is known as the *Tyndall effect* and is not observed with homogeneous liquids. A silver iodide sol, prepared by mixing solutions of silver nitrate and potassium iodide, is a useful system for observing the Tyndall effect. Familiar examples of this effect are the beams from a cinema projector and a car headlight on a foggy night.

The theory of light scattering is beyond the scope of this book but it is useful to consider the effect qualitatively in order to appreciate the kind of information that can be obtained concerning the structure of colloidal solutions from light scattering measurements. Essentially, the incident light interacts with the atoms of the colloidal particles, and those in turn re-emit in all directions the scattered light of the same wavelength.

For particles which are small compared with the wavelength of the incident light and do not absorb in the visible region, the intensity of the scattered light increases appreciably as the wavelength decreases, being inversely proportional to the fourth power of the wavelength. Hence, if the incident light is white, the shorter wavelengths of the visible spectrum are scattered most and the colloidal solution appears to be blue when viewed from the side.

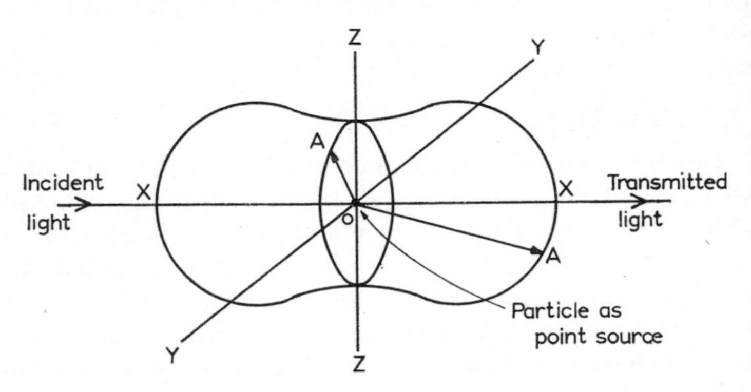

FIG. 21. Light scattering from a small spherical particle with unpolarized incident light.

On this basis Lord Rayleigh was able to explain the blue colour of the sky. The longer wavelengths are least affected by scattering and therefore the transmitted beam has an orange-red colour. This explains the colour of the sunrise and sunset.

Assuming the small particles to behave as point sources of radiation, Rayleigh, during the years 1871–99, showed theoretically that the intensity of scattered light is distributed symmetrically around an axis at 90° to the incident beam. This is illustrated in Fig. 21. The length of the line OA is proportional to the intensity of scattered light in the direction indicated. Furthermore, the total intensity of the scattered light I from small particles is directly proportional to the number of particles and to the second power of their volume. This intensity is given by the Rayleigh equation

$$I = I_0 \cdot \frac{24\pi^3}{\lambda^4}\left(\frac{n_1^2 - n_0^2}{n_1^2 + 2n_0^2}\right)^2 NV^2$$

in which I_0 is the intensity of incident light of wavelength λ (in the medium), N the number of particles per cubic centimetre each of volume V, n_0 the refractive index of the medium and n_1 that of the particle. This equation also shows that the greater the difference between n_1 and n_0, the greater is the intensity of the scattered light.

An important property of light scattered from small particles is that at 90° to the incident (unpolarized) beam it is totally polarized, and partially polarized at all other angles except in the direction of the incident light.

For particles comparable in size with the wavelength of the incident light the intensity diagram is no longer symmetrical, the intensities in the forward direction being greater than those in the backward direction. This arises from interference between the light scattered from different parts of the particle, which can no longer be treated as a point source. Also some depolarization occurs at 90° to the incident beam.

A general theory of light scattering was developed by Mie in 1908 from which it is possible (although laborious) to calculate the intensity of scattered light at any angle as a function of particle size and refractive index. Using these calculations and measurements of intensities at various angles to the direction of the incident light (45°, 90°, 135° are frequently used) together with the depolarization at 90°, it is possible to obtain information on particle number, size and shape.

Colloidal particles are too small to be resolved in an ordinary microscope but can be seen in an ultramicroscope, which makes use of their light-scattering properties. The ultramicroscope consists of an ordinary microscope fitted with a special illuminating system such that the main beam does not enter the objective. The slit ultramicroscope (Fig. 22) of Siedentopf and Zsigmondy

(1903) employs a very intense light beam which projects horizontally an image of a slit into the field of view of the microscope inside a cell containing the colloidal solution. The individual particles scatter light and appear to the observer as bright specks against a dark background (rather like the stars in the night sky). It must be realized that it is not the actual particles that are observed but the light scattered from them and the size of the specks is not simply related to particle size. However, particle concentration can be determined by counting the specks in a

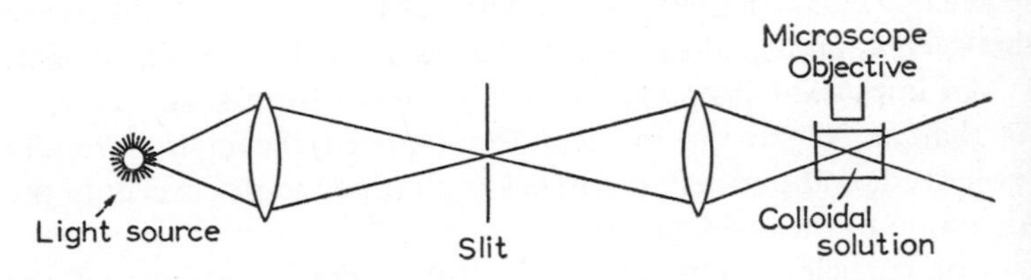

FIG. 22. Schematic diagram of the slit ultramiscroscope.

known volume of liquid and knowing the total amount of substance in that volume the average particle size can be calculated. In an aqueous gold sol particles as small as 20 Å radius are visible in the ultramicroscope because of the large difference in refractive index between particle and medium and therefore high scattering intensity. Large molecules such as proteins in water are little different in refractive index from the surrounding medium and are not resolved in the ultramicroscope. Use is made of the ultramicroscope in detecting the presence of individual particles and for the microelectrophoresis technique with small particles. Dark ground condensers which may be fitted beneath the stage of an ordinary microscope serve the same purpose as the slit arrangement, and are now available from microscope manufacturers.

The electron microscope, with a limit of resolution which is considerably lower than that of an ordinary microscope, provides a useful method for the study of size and shape of colloidal particles. Modern electron microscopes can resolve particles as small as 10 Å and therefore cover the whole colloidal range.

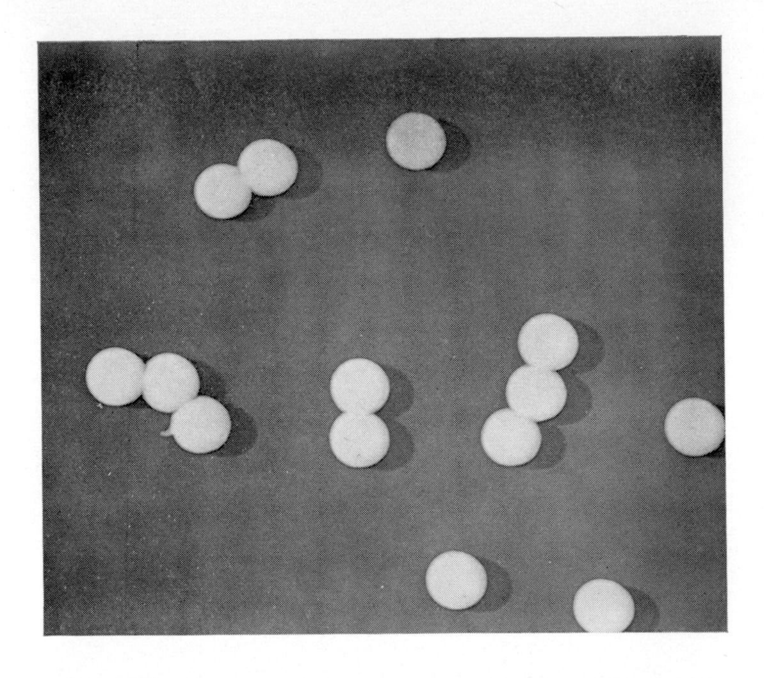

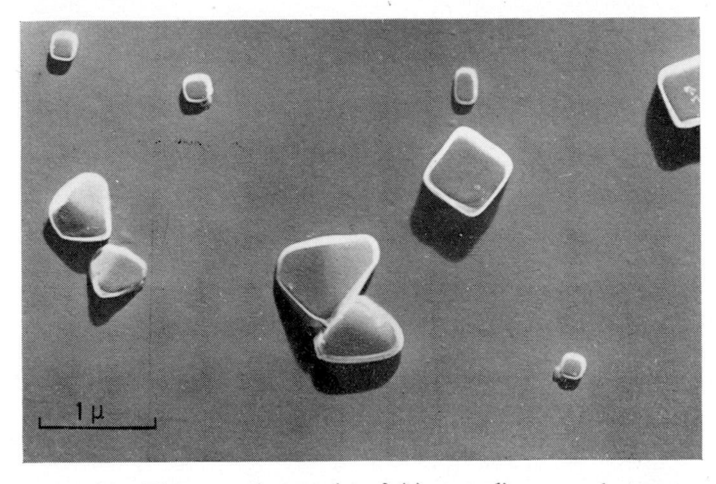

FIG. 23. Electron micrographs of (*a*) monodisperse polystyrene lattices of radius 2650 ± 50 Å, and (*b*) polydisperse silver chloride particles (carbon replica).

Some electron micrographs of colloidal particles are shown in Fig. 23. The shadows, which are produced by a special technique, give an excellent indication of particle shape.

Brownian Movement

Colloidal particles observed in the ultramicroscope are found to exhibit continuous, haphazard motion in all directions. This was first noticed by the botanist Robert Brown in 1827 with pollen grains suspended in water, and is known as the *Brownian Movement.* The explanation, presented about 40 years later, is that the motion is caused by irregular bombardment of the particles by the molecules of the liquid in which they are suspended. Einstein in 1905 showed theoretically that the average displacement x along any given axis in time t of a particle or radius a in a medium of viscosity η is given by

$$x^2 = \mathbf{R}Tt/3\pi N_0 \eta a$$

where $\mathbf{R}$ is the gas constant, T the absolute temperature and N_0 Avogadro's number. The equation was confirmed experimentally by Perrin in 1909 and others using a variety of suspensions in the ultramicroscope and from measurements of the average displacement values for Avogadro's number in the region of 6×10^{23} were found. These results show excellent agreement with the present value of $6{\cdot}023 \times 10^{23}$ and constitute convincing evidence for the validity of the kinetic theory, which assumes that the molecules of any gas or liquid are in continuous motion.

Origin of the Charge on Lyophobic Colloidal Particles

The majority of lyophobic sols, particularly in aqueous solution, owe their stability to the fact that the particles are charged. Only in comparatively few cases is the origin of this charge completely understood.

A great deal of fundamental work has been carried out on the silver halides. A silver iodide sol, for example, can be prepared by mixing aqueous solutions of silver nitrate and potassium iodide at appropriate concentrations. If the silver nitrate is in sufficient excess the particles are positively charged and move towards the cathode in electrophoresis, while in excess potassium iodide the particles carry a negative charge. The charge arises from an excess of either silver or iodide ions on the particle, and since the particle is in equilibrium with these ions in solution the magnitude of the charge and surface potential depends on their concentration in solution. The solubility product of silver iodide in water is approximately 10^{-16}, and therefore small particles of silver iodide exist in a saturated solution in which the product of the concentrations of silver and iodide ions is 10^{-16}. Iodide ions are more strongly held in the crystal lattice than silver ions and the latter will escape more readily into the solution. Since the particle charge is a result of a dynamic equilibrium between ions leaving the surface by thermal agitation and those depositing on it from solution, the concentration of silver ions in solution must be greater than that of the iodide ions for the particle to be neutral. This occurs when the silver and iodide ion concentrations are approximately 10^{-6} and 10^{-10} g ion l^{-1} respectively and is known as the *zero point of charge*. At higher silver ion concentrations (and lower iodide) the particle is positively charged, and has a negative charge at higher iodide (and lower silver) concentrations. The silver and iodide ions are called the *potential-determining* ions.

In some cases the charge arises from dissociation of a surface compound into ions (ionization) when the solid particle is brought into contact with water. A surface compound AB might dissociate into A^+ and B^- ions, the former remaining attached to the particle giving it a positive charge while the latter are set free into the liquid phase forming the diffuse layer associated with the charged particle. This mechanism has been proposed for gold, silver, sulphur and ferric hydroxide sols although the nature of the dissociating compounds has not been well characterized.

Another possibility is that ions from solution, other than those that form the crystal lattice, are preferentially adsorbed on the surface. This has already been suggested as a mechanism for electro-osmosis. Particles of paraffin wax may become negatively charged in pure water as a result of preferential adsorption of hydroxyl ions. Long-chain organic ions adsorb on carbon black particles from aqueous solution as in the case of the dodecyl sulphate ion ($C_{12}H_{25}SO_4^-$) from the detergent sodium dodecyl sulphate, and the particle acquires a negative charge.

Stability of Lyophobic Sols

An understanding of the factors that control the stability of colloidal dispersions is of considerable importance. There are many practical examples of dispersions, such as paints and inks, for which a high degree of stability is required. On the other hand stable dispersions can sometimes be a nuisance, as in an industrial process where separation of the dispersed material from the medium is desired, and therefore a convenient way must be found to effect precipitation of the colloid.

A colloidal dispersion may be considered as stable if there is no change in the particle number and size with time. In the strict thermodynamic sense the system is never really stable because the large interfacial free energy associated with the small particles would tend to decrease by recrystallization to a value for which the surface area is as small as possible, that is when the particles have combined to form one large crystal. The rate at which this occurs is usually insignificant. Sedimentation under gravity would also reduce the number of dispersed particles but in colloid systems the particles are so small that they settle very slowly and thermal agitation is sufficient to keep them dispersed. A very high-speed centrifuge would be necessary to give a measurable sedimentation rate.

The primary cause of instability is flocculation, which is the sticking together of particles into loose clusters that settle quite rapidly. In the clusters the particles retain their original identity. The frequency of collisions and thus the rate of flocculation is determined by such factors as Brownian motion, particle size and concentration, and the attractive and repulsive forces that exist between particles.

It has been known for some time that the addition of electrolytes to a stable lyophobic sol can lead to flocculation. Early workers in the field compared the changes in electrophoretic mobility on addition of electrolyte with the stability of the sol, and found that flocculation corresponded to a reduction in the zeta potential. It was then realized that the stability was connected in some way with the Coulombic repulsion between the charged particles. Values of the concentrations of various inorganic electrolytes that are sufficient to flocculate a sol (the flocculation values) have been reported for a number of both positively and negatively charged systems. They show that the flocculating power of an electrolyte depends upon the valency of the ion of opposite charge to that of the particle and the effect increases markedly with the valency. The valency of the ion of the same sign is of little importance. This relationship between valency and stability is known as the *Schulze–Hardy rule.* Table 5 gives some results for the flocculation of a negative silver iodide sol and a positive aluminium oxide sol.

From these experiments it was thought that flocculation by an electrolyte is due to adsorption of the ion of opposite sign on the

TABLE 5

FLOCCULATION VALUES (IN MILLIMOLE/LITRE)

Silver iodide (−)		Aluminium oxide (+)	
Electrolyte	Flocculation value	Electrolyte	Flocculation value
$LiNO_3$	165	$NaCl$	43·5
$NaNO_3$	140	KCl	46·0
KNO_3	136	KNO_3	60·0
$RbNO_3$	126	K_2SO_4	0·30
$Mg(NO_3)_2$	2·60	K_2CrO_4	0·95
$Ca(NO_3)_2$	2·40	$K_2Cr_2O_7$	0·63
$Sr(NO_3)_2$	2·38	$K_3Fe(CN)_6$	0·080
$Al(NO_3)_3$	0·067		
$La(NO_3)_3$	0·069		

particle surface thus reducing the net charge on the particle and the zeta potential. There are, however, a number of aspects of flocculation that are not explained by this simple theory although the general idea that the stability is governed by the charge forms the basis of the modern theoretical approach. This will now be considered.

When two charged colloidal particles approach each other there are principally two forces that exist between them, namely the van der Waals attraction and the repulsion associated with the charged particles and their double layers.

The van der Waals attractive forces between atoms and molecules, the existence of which provides an explanation of such phenomena as the liquefaction of inert gases, have only a very short range of the order of atomic dimensions. For colloidal particles which contain a large number of atoms or molecules, the forces are assumed to be additive, which results in their having a much longer range, comparable with the size of the particle. The attractive potential energy V_A for small distance between two spherical particles of radius a, when the distance between the surfaces is b, is given by

$$V_A = -Aa/12b$$

where A is a constant whose value depends on the nature of the material of the particles and the medium separating them. The attractive force is independent of the charge on the particle.

It is useful at this stage to consider the rate at which a sol would flocculate if no repulsion existed between the particles. This was considered by Smoluchowski in 1916 who proposed a theory of rapid flocculation based on the idea that particles will collide through their Brownian motion and that every collision leads to a permanent contact removing one particle from the system. The rate of flocculation, i.e. the rate of decrease of the total number of particles, depends on the size and concentration of the spherical particles and is given by

$$\text{rate of flocculation} = 4\pi DrN^2$$

in which D is the diffusion coefficient of the particles in the medium (a function of particle size, viscosity of the medium and temperature), r the collision radius (twice the particle radius), and N the number of particles per cubic centimetre. Using Einstein's equation $D = \mathbf{k}T/6\pi\eta a$, where $\mathbf{k}$ is the Boltzmann constant, T the absolute temperature, η the viscosity of the medium and a the particle radius, it follows that

$$\text{rate of flocculation} = 4\mathbf{k}TN^2/3\eta \tag{10}$$

The time ($t_{\frac{1}{2}}$) in which the number of particles is reduced to half its initial value is N^0 a useful measure of the rate of flocculation and is obtained by integration of eqn. (10) giving

$$t_{\frac{1}{2}} = 3\eta/4\mathbf{k}TN^0$$

For water as the dispersion medium at 25°C

$$t_{\frac{1}{2}} \sim 2 \times 10^{11}/N^0 \text{ sec}$$

Aqueous silver halide sols, prepared by mixing solutions of silver nitrate and the appropriate inorganic halide, have particle concentrations of the order of 10^{10}–10^{12} cm^{-3}. Hence, under these conditions flocculation would be very rapid. The observed stability of these sols must, therefore, be associated with the particle charge.

Since a charged particle with its double layer is electrically neutral, no net Coulombic force exists between particles at large distances from each other. As the particles approach their diffuse double layers interpenetrate, and this gives rise to a repulsive force which increases as the particles get closer together. The distance between the particles at which this repulsion becomes important increases with the thickness of the double layer. The repulsive force increases with the surface potential. Stability, therefore, will be determined by both the charge on the particle and the thickness of the double layer. To prevent flocculation the force of repulsion must be appreciably greater than that of attraction which also becomes important at short distances of approach.

The behaviour of lyophobic sols is, therefore, determined by the relative magnitude of the forces of attraction and repulsion. A useful method of representing the interaction between two colloidal particles is a plot of the potential energies as a function of their distance apart. This is illustrated in Fig. 24(*a*) for a stable system; the solid curve is the resultant of the curves for attraction and repulsion and thus represents the total potential energy as a

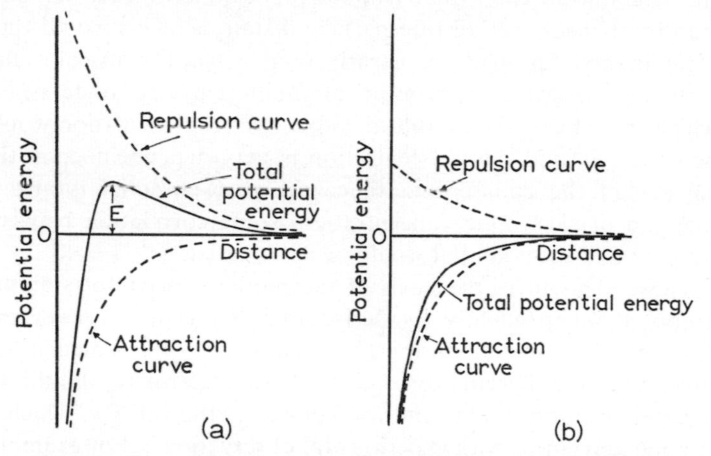

FIG. 24. Potential energy versus distance between particles for (*a*) a stable sol, and (*b*) after addition of sufficient electrolyte to cause flocculation.

function of distance. For very small distances the attraction predominates, but at intermediate distances repulsion becomes significant and gives rise to a maximum in the total potential energy curve, which constitutes an energy barrier to flocculation. Particles must possess energy in excess of E to approach close enough for them to stick together.

The effect of addition of electrolytes can now be understood. The simplest case is that in which there is no specific interaction between the electrolyte and the particle, as, for example, in the addition of various nitrates (except silver nitrate) to a negative silver iodide sol. The surface potential is determined by the concentrations of silver and iodide ions and the added nitrate only affects the nature of the double layer. Increasing the concentration of the so-called *indifferent electrolyte* causes a compression of the double layer, and therefore a decrease in the distance from the particle at which repulsion begins to occur. As mentioned earlier, this effect increases markedly with the valency of the ion of opposite sign to that of the particle. The net result is a decrease in the height of the energy barrier and, as shown in Fig. 24(*b*), it may lead to attraction at all interparticle distances. This would then result in rapid flocculation. Clearly, there will bc for any particular system a critical concentration of indifferent electrolyte below which the sol would be stable and above which flocculation would be observed, and this concentration is very dependent upon the valency of the counterion. Since the zeta potential is also a function of electrolyte concentration, some correlation between stability and the potential would be anticipated.

These ideas form the basis of the modern theory of stability proposed, independently, by Derjaguin in Russia and Verwey and Overbeek in Holland during the period 1938–48. Calculations show that the flocculation value of an electrolyte should be inversely proportional to the sixth power of the valency, which is in good agreement with experimental observations. For example, the mean flocculation values of mono-, di-, and trivalent cations for a negative silver iodide sol are 142, 2·43 and 0·068 millimolar respectively, which are in the ratio 1 : 0·017 : 0·0005. The theoretical ratio is

$$\frac{1}{(1)^6} : \frac{1}{(2)^6} : \frac{1}{(3)^6} \quad \text{or} \quad 1:0{\cdot}016:0{\cdot}0013$$

The theory thus supports the Schulze–Hardy rule.

The Preparation of Lyophobic Sols

The methods available for the preparation of a dispersion of an insoluble substance in a liquid medium, with particles in the colloidal range (10–10,000 Å) may be conveniently divided into two types: (a) *dispersion methods* in which coarse particles or aggregates are broken down to colloidal dimensions, or (b) *condensation methods* involving the aggregation of atoms, ions or molecules into colloidal particles.

(a) *Dispersion methods.* Solids may be disintegrated into small particles by mechanical means. On a laboratory scale, grinding the material in an agate mortar usually produces some particles in the colloid range, although the majority are somewhat coarser. Although the smaller particles have a tendency to aggregate by virtue of their large surface free energy, this may be overcome by the addition of a surface active agent, such as a long-chain detergent, the ions of which adsorb on the surface and give the particles a charge. In the colloid mills used in industry, a mixture of the coarse material and dispersion medium is subjected to intense shearing forces. Another method which is quite effective for laboratory purposes, although not yet applied commercially to any great extent, involves irradiation of the aggregated material by ultrasonic waves. Its application is limited because ultrasonic waves can, in certain circumstances, cause flocculation.

The redispersion of precipitates (which must usually be freshly prepared) is known as *peptization.* For example, freshly precipitated silver iodide can be peptized by washing, followed if necessary by the addition of dilute solutions of silver nitrate or potassium iodide. The precipitate consists of colloidal silver iodide particles which may have flocculated because of the presence of a high concentration of electrolyte. Washing removes excess electrolyte and this alone may cause the particles to redisperse, provided the surface potential is large enough for stability. If not, the addition of potential determining ions (Ag^+ or I^-) would ensure the formation of a stable sol.

(b) *Condensation methods.* All the techniques that are available for the precipitation of a solid from homogeneous solution may be used for the preparation of lyophobic sols. With the proper choice of conditions the aggregation of single atoms, ions or molecules will lead to the formation of particles of colloidal dimensions rather than large crystals.

The first step in the precipitation process is the formation of a supersaturated solution, which can readily be prepared by a variety of methods such as decrease in solubility and chemical reaction. Sulphur sols may be obtained by pouring an alcoholic solution of sulphur into water in which the sulphur is much less soluble, or by the reaction between an acid and sodium thiosulphate. Double decomposition reactions are quite common as in

$$AgNO_3 + KI \rightarrow AgI\downarrow + KNO_3$$
$$BaCl_2 + Na_2SO_4 \rightarrow BaSO_4\downarrow + 2NaCl$$

A solution which is only slightly supersaturated may be stable for an indefinite period of time, but on increasing the concentration a point is reached when the formation of the solid phase occurs spontaneously. Initially the atoms, ions or molecules cluster together to form very small units known as *nuclei*, whose existence is only possible when a certain degree of supersaturation has been reached. This concentration is determined by the solubility of the nucleus which can be shown to be much larger than that of the bulk solid. When the solution is only slightly supersaturated (with respect to the solubility of the bulk solid) the nucleus would have to be quite large for it to be stable and the probability of its formation is very small. Increasing the degree of supersaturation corresponds to a decrease in the size of the stable nucleus, and the much smaller nucleus has a greater probability of formation. Spontaneous formation of nuclei can, therefore, only be expected with a relatively large supersaturation. Once the nuclei are formed they then grow further by assimilating more material from solution until the supersaturation is relieved. The ultimate size of the crystals will depend on the number of nuclei formed and the total amount of material to be precipitated.

The formation of a colloidal solution is favoured by the production of a large number of nuclei on which growth is relatively slow. If the nuclei are formed over a short period of time and they grow at a uniform rate, the final particles will all be approximately the same size and the system is called *monodisperse*.

Very carefully controlled conditions are necessary for the formation of monodisperse sols. In most cases nucleation and growth are occurring simultaneously and the final particles originate from nuclei that are formed at different times. This is the case in the mixing of silver nitrate and potassium iodide solutions when the final particle size range is quite wide, and a *polydisperse* silver iodide sol is obtained.

Monodisperse sulphur sols have been successfully prepared by La Mer by the acid decomposition of sodium thiosulphate in aqueous solution. Low concentrations of reactants ($\sim$0·001 M) are used and the sulphur is produced very slowly over a period of an hour or so, while the supersaturation builds up to quite a high level. At the critical degree of supersaturation a large number of nuclei ($\sim 10^6$ cm^{-3}) appear spontaneously and grow at such a rate as to quickly relieve the supersaturation to a level that further nucleation is prevented. Sulphur continues to be generated by the reaction but slowly enough so that the degree of supersaturation is kept low. Hence, a monodisperse system of growing particles is obtained and the growth can be stopped at any time by addition of potassium iodide which reacts with the remaining thiosulphate.

It has been assumed, so far, that the sol particles do not flocculate since this would affect the final state of the sol. However, in many cases quite a high concentration of electrolyte may be present in solution and this could lead to instability. For example, potassium nitrate is a by-product of the reaction between silver nitrate and a potassium halide, and stable silver halide sols are only produced when the electrolyte concentration is kept low. Extraneous electrolyte may conveniently be removed by *dialysis*. The simplest method is to pour the sol into a cellophane bag and place the bag in a container of distilled water. The ions being small diffuse through the pores of the cellophane while the larger colloidal particles remain in the bag. Frequent renewal of the water on the outside helps to increase the rate of dialysis but it is, at best, a very slow process. Dialysis can be accelerated by applying a potential difference across the membrane, the ions migrating faster under the influence of the applied field. This is called *electrodialysis*. It is usual not to remove all the electrolyte since, as in the silver halides, a certain concentration of potential determining ions is required to maintain sufficient charge on the particles for stability.

LYOPHILIC SOLS

The term *lyophilic* is used for colloidal systems in which the dispersed material has a high affinity for the dispersion medium in contrast with the lyophobic systems for which the affinity is small. It is for this reason that lyophilic systems usually form spontaneously when the material and medium are brought together. Examples of lyophilic colloids are molecules of high molecular weight such as the synthetic polymers (polystyrene, nylon, synthetic rubber), the naturally occurring proteins (egg albumen, haemoglobin), gum, resins and natural rubber. When dispersed in a liquid medium the *particles* are largely separate molecules whose dimensions are in the colloid range. The behaviour of these lyophilic dispersions varies widely according to the concentration of the colloid and the shape of the molecules. At very high concentrations of long thread-like molecules such as gelatin in water, the system may be quite rigid and consists of a mass of entangled gelatin molecules with water trapped between them. When the particle concentration is low they do not differ essentially from ordinary molecular solutions, except in the shape of the solute molecule. The properties of relatively dilute lyophilic sols will be considered here with special reference to solutions of polymers and proteins.

Polymer and protein molecules contain a large number of smaller molecular units which are joined together through primary valencies in a more or less repetitive manner. They are called *macromolecules*. Polyethylene (polythene), for example, contains many ethylene molecules which are linked end-to-end to form long linear macromolecules:

$$n(\mathrm{CH_2{=}CH_2}) \rightarrow \mathrm{—CH_2—CH_2—CH_2—CH_2—}$$

Another well-known polymer is nylon which results from the condensation of hexamethylenediamine and adipic acid with the elimination of water:

$$n\mathrm{H_2N(CH_2)_6NH_2} + n\mathrm{HOOC(CH_2)_4COOH} \rightarrow$$
$$\mathrm{H[—HN(CH_2)_6NHOC(CH_2)_4CO—]_{\mathit{n}}—OH} + (n-1)\mathrm{H_2O}$$

Rubber is a naturally occurring hydrocarbon polymer. Polymer molecules such as polyethylene and nylon are usually linear and in solution are very flexible. As a result they can exist in various configurations depending on the solvent, ranging from spheres when completely coiled up to rods when stretched out.

The naturally occurring proteins are very complex molecules built up by the condensation of a large number of amino-acids of the type

$$\begin{matrix} \mathrm{R} & & \mathrm{NH_2} \\ & \mathrm{C} & \\ \mathrm{H} & & \mathrm{COOH} \end{matrix}$$

where R is a side chain that may be only hydrocarbon or may also contain amino or carboxyl groups. In most proteins the amino-acids are joined together to form long chains in the following manner:

$$\mathrm{NH_2—CH(R)—CO—NH—CH(R)—CO—NH—CH(R)—CO—\ldots\ldots—NH—CH(R)—COOH}$$

Free amino and carboxyl groups occur at the ends of the protein chain and, in certain cases, at various points along the chain. The presence of these groups means that in aqueous solution a protein can function both as an acid and as a base depending on the pH, and it is this property that determines the electrical behaviour of protein sols.

The study of the behaviour of polymer and protein solutions constitutes a very important part of modern science. Biochemists have long recognized the role played by proteins in living processes and important in their research is a knowledge of the weight, shape and electrical polarity of the protein molecules. Polymeric materials, both natural and synthetic, are very common in everyday life and their properties are to a large extent governed by molecular size and shape. Very useful information may be obtained from a study of the colloidal properties of polymer and protein molecules in solution. These will now be considered and comparisons made with lyophobic sols where appropriate.

Electrical behaviour

Evidence for the electrical charge of lyophilic colloids in aqueous solution is provided by the fact that they conduct electricity and show electrophoresis. The charge arises from dissociation of certain groups in the molecule, such as —COOH, —SO_3H and —NH_2. Since proteins contain both amino and carboxyl groups which can ionize in solution, the resultant charge on the molecules depends on the number of these groups present and their dissociation constants. Addition of acid to an aqueous protein solution suppresses the dissociation of the carboxyl groups but the number of —NH_3^+ groups is increased, so that the molecules become more positive. The reverse effect is produced when a base is added, the charge becoming more negative. For each protein there is one value of pH for which the numbers of carboxyl and amino ions on the surface of the molecules are equal and the resultant charge is zero. This is called the *isoelectric point* and may conveniently be determined by measuring the electrophoretic mobility at various pH values. The pH corresponding to zero mobility is the isoelectric point. Many proteins have their isoelectric points at values of pH in the region of 4 to 5.

The moving boundary method has been developed by Tiselius for investigating the electrophoretic behaviour of colourless protein solutions. The essential part of the Tiselius apparatus is a

U-shaped glass cell of rectangular cross-section, shown in Fig. 25(*a*). It is built in sections which can be moved sideways relative to one another. The lower part of the cell, up to *AB*, is filled with protein solution which is then displaced relative to the upper part [Fig. 25(*b*)]. The top sections are then filled with the dispersion

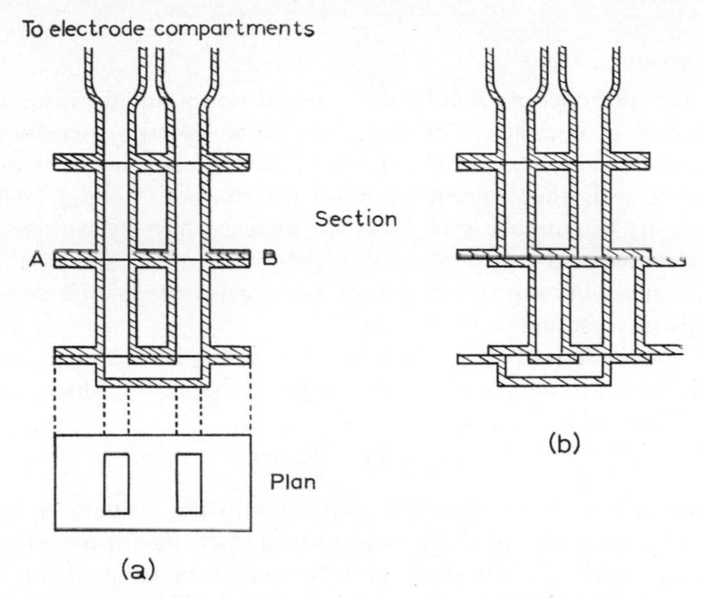

FIG. 25. Tiselius electrophoresis cell. (*a*) Cross-section and plan view of cell in working position. (*b*) One of the centre sections displaced during the filling operation.

medium and the two limbs of the U-tube connected to two large electrode compartments. By sliding the lower parts into position the sections are joined together forming a sharp boundary between the colloidal solution and the medium. The movement of the boundary during electrophoresis is determined by an optical method which depends on the differences in refractive index at the boundary. If the colloidal solution contains a number of substances having different electrophoretic mobilities, separate boundaries will form and from their rates of movement the substances can often be identified. For example, serum has been shown by electrophoresis to contain four different proteins. If the boundaries are sufficiently far apart it is possible to separate the various components, which may be inseparable by chemical methods.

Viscosity

The introduction of colloidal material into a liquid causes an increase in viscosity. For dilute lyophobic sols the viscosity is about the same as that of the medium and increases regularly but slowly with the concentration of the dispersed phase. The viscosity of lyophilic sols, however, increases rapidly with concentration, and the very much higher viscosity at comparable concentrations represents one of the most striking differences between these and lyophobic sols.

Einstein derived an equation for the viscosity of dilute colloidal solutions, assuming the particles to be rigid spheres, which may be expressed in the form:

$$\eta/\eta_0 = \eta_r = 1+2{\cdot}5\phi$$

where η is the viscosity of the sol, η_0 that of the medium, η_r the relative viscosity and ϕ the volume fraction of the disperse phase (the ratio of the volume of the particles to the total volume). From this equation it follows that a plot of η_r against ϕ should be linear with a slope of 2·5, and the viscosity depends only on the total volume of the particles and is independent of particle size. There are numerous examples of experimental measurements on lyophobic sols that have confirmed Einstein's equation for spherical particles at low concentration. However, there are cases of lyophilic systems (aqueous glycogen and egg albumen sols) for which the slope of the $\eta_r - \phi$ plot is greater than 2·5, suggesting that the actual volume fraction is larger than that of the corresponding dry material. An explanation of this difference is that the particles are strongly solvated. Since the effective size in viscosity measurements is that of the particle plus its solvent layer an increase in volume fraction would be expected. Solvent molecules may also penetrate into the interior since these protein particles have a loose structure, and this would also lead to an increase in ϕ.

The majority of lyophilic colloidal solutions do not contain spherical particles, and the Einstein relation is not obeyed. The viscosity of solutions containing linear macromolecules is very much greater than that for spheres under the same conditions, and increases very rapidly with concentration. The viscosity is also dependent on the size of the molecules. Staudinger suggested that a relationship exists between viscosity and molecular weight which may be written as

$$(\eta_{sp}/c)_{c=0} = KM$$

where η_{sp} is the specific viscosity $[(\eta-\eta_0)/\eta_0]$, c the concentration of colloid in grammes per 100 ml of solution, M the molecular weight and K a characteristic constant for the system. The Staudinger law is not now regarded as being generally valid and has been replaced by the equation

$$(\eta_{sp}/c)_{c=0} = KM^{\alpha}$$

in which α is another constant characteristic of the colloid system and has values varying between 0·5 and 1·5. This equation is of great practical importance and is widely used for the determination of molecular weight of polymers, since the viscosity technique is simple and rapid. The viscosity is measured at a number of polymer concentrations and the values of η_{sp}/c extrapolated to $c = 0$ to give the molecular weight. Values of K and α are established from viscosity measurements on samples of the polymer of known molecular weight (determined by another method such as light scattering or osmometry).

The addition of a small quantity of electrolyte to a lyophilic sol containing charged particles causes a large decrease in viscosity. Experiments show that the effect is determined by the ion of opposite charge to that of the sol, and its influence increases markedly with the valency. For many proteins in dilute aqueous solution the viscosity varies according to the pH, with a minimum value at the isoelectric point and considerably higher viscosities at lower and higher values of pH. From these observations it is evident that the changes are associated with the electrical double layer around the particles. The increase in viscosity due to the electrical charge is called the *electroviscous effect*.

Stability

It has already been suggested that lyophobic sols are unstable in the thermodynamic sense and flocculation is prevented by some sort of energy barrier. In aqueous media they are easily flocculated by small quantities of electrolytes which compress the double layer and reduce the energy barrier. Prolonged dialysis usually causes flocculation. Electrolytes have the same effect on lyophilic systems, as indicated by viscosity and electrophoresis measurements, but in most cases no flocculation occurs. They are also stable to prolonged dialysis. It is even possible to obtain protein sols which are quite stable at the isoelectric point, when the net charge on the particles is zero. Solutions of rubber and polystyrene in benzene are also stable and in these cases charge effects are impossible. It is apparent, therefore, that some other factor plays an important part in the stabilization of lyophilic sols.

It is now recognized that the stability of lyophilic colloidal solutions is also associated with the degree of solvation of the particles. This conclusion is chiefly based on the fact that large quantities of electrolytes precipitate aqueous protein sols and that their flocculating power is directly related to their solubility in water and their tendency to become hydrated. At high concentrations it is supposed that added ions compete for the water present and this results in dehydration of the colloidal particles and reduced stability. The effect is called *salting out*. The ions can be arranged in a series of decreasing precipitating power known as the *Hofmeister* or *lyotropic series* as follows sulphate > acetate > chloride > nitrate > bromide > iodide > thiocyanate > magnesium > calcium > barium > lithium > sodium > potassium. The valency of the ions is of much less importance than in the case of lyophobic sols (Schulze–Hardy rule). Ammonium sulphate is frequently used for salting out proteins because it is very soluble in water giving ions which have a high affinity for the solvent.

Lyophilic sols can also be precipitated by the addition of liquids that arc miscible with the dispersion medium but do not dissolve the colloidal material. For example, gelatin dissolves in water but is insoluble in alcohol. Addition of alcohol to an aqueous solution of gelatin causes precipitation, and it is believed that the surface of the colloid becomes dehydrated, giving it typical lyophobic properties. The sol might still be stable if sufficient charge existed on the particles, but relatively small amounts of electrolytes would cause precipitation. Another example is the precipitation of rubber from benzene solution by alcohol. From these observations it is clear that the essential difference between lyophobic and lyophilic sols is in the interaction between the disperse material and the dispersion medium.

DETERMINATION OF MOLECULAR WEIGHT

Both in the study of polymerization and in biological research a knowledge of the size of the molecules is important. The viscosity method for molecular weight determination has already been mentioned. The other methods employed most frequently involve measurement of osmotic pressure, light scattering and sedimentation in the ultracentrifuge, and these will now be discussed briefly.

Molecular Weight by Osmometry

Molecular weights of small molecules are frequently determined in the laboratory by the measurement of colligative properties, such as elevation of the boiling point or depression of the freezing point, which depend only on the number of molecules in solution. However, when these methods are applied to substances of high molecular weight the effect is too small to be measured. The only colligative method applicable to colloidal solutions is that of osmotic pressure measurement. The pressures involved are small and their measurement requires special techniques.

The osmotic pressure π is related to the concentration of solute c (in grammes per litre) and the molecular weight M by the van't Hoff equation

$$\pi/c = RT/M$$

where R is the gas constant and T the absolute temperature. This equation holds for dilute sols containing spherical particles, such as aqueous solutions of egg albumen, but deviations occur with the linear protein and polymer molecules. For these the equation is modified as follows:

$$\pi/c = (RT/M)+Kc$$

where K is a constant characteristic of the system and is independent of molecular weight. A plot of π/c versus c is linear with an intercept at $c = 0$ of RT/M.

Usually the solution contains molecules of varying molecular weight and the value of M obtained is an average value. Since the osmotic pressure depends on the number of molecules present M is called the *number average molecular weight.*

Molecular Weight by Light Scattering

The turbidity τ of a colloidal solution is defined as the fractional decrease in the intensity of the incident light on passing through 1 cm of solution due to scattering by the particles in all directions. For a macromolecular solution it is related to the molecular weight M and the concentration c (in grammes per cubic centimetre) by the Debye equation

$$Hc/\tau = (1/M)+2Bc$$

where B is a constant and H is a complex function containing the optical constants of the system. H is given by the relation

$$H = \frac{32\pi^3 n_0^2}{3N_0 \lambda^4}\left(\frac{dn}{dc}\right)^2$$

where N_0 is Avogadro's number, λ the wavelength of the light (*in vacuo*), n_0 the refractive index of the solvent and dn/dc the variation of refractive index of the solution with concentration.

The turbidity is obtained by measuring the intensity of scattered light at 90° to the incident beam. Since there is only a very small difference between the refractive indices of a polymer or protein solution and the solvent, a special refractometer (called a differential refractometer) is required for the measurement of dn/dc.

To determine the molecular weight of the protein or polymer, τ and dn/dc are obtained for several dilute solutions, and the values of Hc/τ are plotted against c. The intercept at $c = 0$ of the best straight line drawn through the points is given by $1/M$.

Since the turbidity depends on both the number and volume of the particles the value of M obtained from light scattering measurements is the *weight average molecular weight*. When the distribution of molecular weights in the solution is very narrow, the number average and weight average molecular weights are almost identical.

Molecular Weight by Sedimentation

The rate at which particles settle under gravity in a liquid medium depends on their size and density and the viscosity of the medium. Particles of diameter greater than 1 μ will usually settle quite rapidly and from observations of the sedimentation rate it is possible to determine the particle dimensions.

Sedimentation of colloidal particles, however, is an extremely slow process but the rate can be increased by subjecting them to a strong centrifugal field in a very high-speed centrifuge, known as an *ultracentrifuge*. With this instrument speeds of the order of 100,000 rev/min are attained and the force on the particle is about a million times that due to gravity.

In a modern ultracentrifuge the transparent cell containing the colloidal solution is placed in a specially designed rotor which is driven by an electric motor. If the particles are of uniform weight they will all move at the same rate away from the axis of rotation,

A sharp boundary then arises between the sedimenting particles and pure solvent across which there will be a rapid change in refractive index. By means of windows in the rotor a beam of light is passed through the cell while it is spinning, and a special optical system is used to follow the movement of the boundary. The molecular weight M is related to the sedimentation rate by the following equation:

$$M = \frac{RT \log_e(x_2/x_1)}{D(1-V\rho)(t_2-t_1)\omega^2}$$

where x_2 and x_1 are the distances the boundary moves in times t_2 and t_1 respectively, ω the angular velocity of the rotor, ρ the density of the dispersion medium, D the diffusion coefficient, V the partial specific volume of the particles (the volume in cubic centimetres occupied by 1 g of the substance), $\boldsymbol{R}$ the gas constant and T the absolute temperature.

The optical system is so designed that the rate of sedimentation of particles of different molecular weight in a polydisperse system can be measured and the individual molecular weights determined. The ultracentrifuge is widely used with biological systems both for molecular weight determination and for the separation of various components.

COLLOIDAL ELECTROLYTES

Another important class of lyophilic colloids is the so-called *colloidal electrolytes*, whose behaviour differs sufficiently from that of polymers and proteins to merit a separate discussion. These are substances which in very dilute solution behave as normal electrolytes but show colloidal properties at higher concentrations. The soaps (alkali metal salts of long-chain fatty acids) and the synthetic detergents (e.g. sodium dodecyl sulphate and cetyl trimethyl ammonium bromide) are the principal members of this class. Others include certain dyestuffs (e.g. methylene blue) and a number of substances of biological importance (e.g. lecithin). The common feature of these substances is that they contain a large non-polar hydrocarbon portion R and an ionizing polar group (e.g. $R—SO_3^-Na^+$, $R—SO_4^-Na^+$, $R—COO^-Na^+$, $R—N^+(CH_3)_3Br^-$). Although they are not electrolytes, the modern commercial non-ionic detergents also possess this polar–nonpolar character and show similar colloidal behaviour to the substances mentioned above. The most important of these are the polyethylene–oxide condensation products such as $R—C_6H_4—O—CH_2CH_2(OCH_2CH_2)_n—OCH_2CH_2OH$, where R is a hydrocarbon chain. A considerable amount of fundamental work has been carried out on the physical properties of aqueous solutions of colloidal electrolytes, in particular the soaps and detergents.

In dilute aqueous solution a typical colloidal electrolyte such as sodium dodecyl sulphate behaves as a normal strong electrolyte. The osmotic pressure increases with concentration as would be expected for a simple electrolyte and the particle weight derived from these measurements is equal to half the empirical weight, corresponding to the presence of sodium and dodecyl sulphate ions in solution. There is, however, a certain concentration (0·008 M for sodium dodecyl sulphate) above which the osmotic pressure shows only a very slow increase, and in this region the solution is found to contain particles which are much larger than the simple molecules. A sudden increase in light scattering also occurs at this concentration and the appearance of the Tyndall effect indicates that the system has become colloidal in nature. The explanation, first put forward by McBain in 1913, is that the colloidal electrolyte aggregates into fairly large units known as *micelles*, and the concentration at which this occurs is the *critical micelle concentration* (C.M.C.). Sharp changes in other physical properties such as electrical conductivity, surface tension, density and solubility, also occur at the C.M.C.

It is now well established that the micelles in solutions of concentration less than about 1 per cent by weight are approximately spherical in shape, as illustrated in Fig. 26 for an electrolyte containing a long-chain anion and a small cation (such as sodium dodecyl sulphate $C_{12}H_{25}SO_4Na$). Other shapes have been suggested for more concentrated solutions but these will not be considered here. In the spherical micelle the hydrocarbon chains are all directed inwards towards the centre leaving the ionized polar groups at the surface. An important factor in micelle formation is the high interfacial energy between hydrocarbon and water which is significantly reduced when aggregation occurs. This is opposed by the electrostatic repulsion between the surface

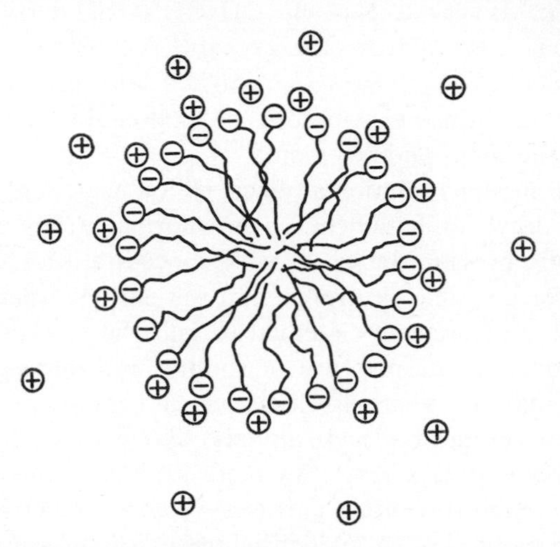

FIG. 26. Schematic representation of a spherical micelle showing the hydrophilic group in the outer part of the micelle and the hydrophobic portion filling the interior.

ions and by thermal agitation. Micelle formation will, therefore, be assisted by decreasing temperature and increasing chain-length, which both cause a lowering of the C.M.C. It may seem quite surprising that aggregation into micelles occurs at a well-defined concentration but application of the law of mass action to the process shows that this is in fact the case.

The micelles carry a high net charge but this is not as large as would be expected if all the electrolyte were dissociated into ions, indicating that some of the counterions remain associated with the micelle. There are about 40 sodium ions associated with the 60 or so long-chain ions in the sodium dodecyl sulphate micelle, the remainder forming the diffuse part of the double layer around the micelle (Fig. 26). Also in solution in dynamic equilibrium with the micelles are a certain number of simple dodecyl sulphate and sodium ions. Addition of further material above the C.M.C. merely increases the number of micelles, the concentration of simple long-chain ions remaining practically constant. Since the non-ionic detergents also form micelles, the possession of a net charge is not essential for stability.

CHAPTER 6

EMULSIONS, FOAMS, AEROSOLS AND GELS

EMULSIONS

The term *emulsion* refers to a disperse system containing two immiscible liquids, one of which is dispersed as small droplets in the other as the dispersion medium. Water (or an aqueous solution) is usually one of the liquids and the other is an organic compound (oil) which is insoluble in the aqueous phase. There are two types of emulsion, oil-in-water and water-in-oil, according to whether the aqueous or the oil phase is the continuous one. These are frequently referred to as O/W and W/O emulsions respectively. In all cases a third component, known as the emulsifying agent, is essential for stability. Familiar examples of emulsions are milk, cream, butter, margarine, mayonnaise, salad cream, cosmetic creams and lotions, bituminous emulsions used in road-surfacing and a variety of agricultural sprays. Emulsions are also widely used in pharmacy.

Emulsion droplets have diameters in the range 1–50 μ and are therefore larger than those found in colloidal solutions. They are usually visible in the ordinary microscope. In the majority of common emulsions the concentration of disperse phase is much greater than that of sols, e.g. cream contains about 20 per cent of fat droplets dispersed in an aqueous medium. Assuming that the droplets are spherical and of equal size, a simple geometrical calculation shows that the maximum volume of one liquid which can be dispersed in another is 74 per cent of the total available volume and is independent of droplet size. If this concentration is exceeded one might expect inversion to occur, the disperse phase becoming the dispersion medium, e.g. a change from *O/W* to *W/O* type emulsion. However, emulsions of both types of much higher concentration are known and this is mainly due to the fact that the emulsifying agent allows the droplets to become distorted from their spherical shape without coalescing. In emulsions of non-uniform droplet size there is also the possibility of introducing small droplets between the larger ones. Stable dilute dispersions of oil in an aqueous medium can be prepared in the absence of an emulsifying agent if the oil occupies only a very small fraction (about 0·1 per cent) of the total volume. Their electrophoretic behaviour and stability to electrolytes closely resemble those of lyophobic sols and such systems are usually regarded as oil hydrosols rather than as emulsions.

It is relatively easy to determine whether a given emulsion is of the *O/W* or *W/O* type. Addition of water to the former results in rapid dilution of the emulsion but would form another layer in the *W/O* case. Also, the conductivity of an *O/W* emulsion should be high, and of the *W/O* type, low. Another method is to add to the emulsion a dye which is only soluble in one of the phases: an oil-soluble dye will readily colour a *W/O* emulsion and conversely the *O/W* type is coloured by water-soluble dyes.

Preparation of Emulsions

An emulsion can be prepared by simply shaking two immiscible liquids together, but unless an emulsifying agent is present the emulsion is unstable and the liquids separate into two layers. Emulsifying agents are substances which are strongly adsorbed at the oil–water interface thereby lowering the interfacial tension. Since in the formation of an emulsion a very large surface area is created between the two liquids, a reduction in the interfacial tension facilitates the formation of droplets and also reduces their tendency to coalesce. However, a substance which markedly lowers the interfacial tension does not necessarily give rise to a stable emulsion. The concentration of the emulsifying agent in a stable emulsion is usually of the order of 1–5 per cent.

Emulsifying agents are numerous. *O/W* emulsions are promoted by many hydrophilic colloids such as proteins, gums and most ionic soaps and detergents. Long-chain alcohols and esters and the oleates and stearates of aluminium, calcium, and magnesium are used in the formation of *W/O* emulsions. Certain finely divided solids which are preferentially wetted by one of the liquids are also capable of acting as emulsifying agents, e.g. various carbon blacks promote *W/O* emulsions and clays give the *O/W* type.

A variety of mechanical devices are available commercially for the preparation of emulsions both on the laboratory and industrial scales. Those which involve high-speed stirring or other agitation methods usually produce emulsions containing large droplets with a range of particle size. Subjecting a coarse emulsion to homogenization results in the production of a smaller and more uniform particle size. In the commercial homogenizer the coarse emulsion is forced at high pressure through a valve operated by a strong spring which allows only a very small clearance. The fat droplets in milk can be reduced from about 10 to 0·5 μ by homogenization and the resulting emulsion is then quite stable, showing little tendency for the cream to separate.

Emulsion Stability

A number of theories have been proposed for the stabilizing effect of the emulsifying agent. Although a reduction in interfacial tension by adsorption of the agent at the oil–water interface facilitates the formation of an emulsion, the fact that alkali metal soaps tend to give the *O/W* type and the soaps of the alkaline earth metals the *W/O* type, even though the interfacial tension lowerings are similar, indicates that other factors are involved in emulsion formation and stability.

The stability of an emulsion is determined essentially by the nature of the interfacial film. The water soluble proteins are very effective emulsifying agents giving *O/W* type emulsions. When an oil is brought into contact with an aqueous protein solution a film of the protein, which is multimolecular and rather viscous, spontaneously forms at the *O*/*W* interface. The oil droplets in an *O*/*W* emulsion are stabilized by the mechanical protection given by the protein film. In milk the fat droplets are stabilized by the protein casein.

Electrical effects are important in the stabilization of both *O*/*W* and *W*/*O* type emulsions when ionic soaps and detergents are used as emulsifying agents. The interfacial film is monomolecular in thickness with the non-polar part of the agent in the oil phase and the polar group in the water phase. The droplets are charged by the ionization of the surface groups and in the *O*/*W* type emulsion the situation resembles that already described for lyophobic sols, e.g. flocculation rates increase with electrolyte concentrations.

The emulsifying effect of a finely divided powder depends on how it is wetted by the liquids. For a solid powder to be an efficient stabilizer it must be taken up at the interface and this will not occur if the solid is completely wetted by either of the liquids since then the particles would remain in one or both of the bulk phases. The contact angle θ between a solid particle and the two liquid phases must, therefore, be finite ($0° < \theta < 180°$), and the type of emulsion formed is determined by which phase preferentially wets the solid. The solid particle will be in a position in which as much as possible of its surface is in the liquid phase which most nearly wets it, and more particles can be accommodated at the interface if this liquid constitutes the outer phase. Carbon blacks, which are preferentially wetted by the oil, promote *W*/*O* emulsions with the bulk of the solid particles in the oil phase, as illustrated in Fig. 27. The layer of solid particles around the droplet gives it mechanical strength and thus confers stability on the emulsion.

Since the majority of oils are lighter than water, the oil droplets in an emulsion, if larger than about 1 μ in radius, rise to the surface under the influence of gravity. This phenomenon is known as *creaming* and can be reduced by addition of substances (e.g. glycerin, gums or gelatin) which increase the viscosity of the continuous phase. Reduction in droplet size by homogenization also slows down creaming. Sometimes creaming is desirable, e.g. in the concentration of latex emulsion to remove the rubber particles, and this may be facilitated by centrifugation or by adding a substance which increases the density of the continuous phase. In creaming, the emulsion droplets flocculate at first to give clusters with a consequent increase in sedimentation rates. Coalescence of the droplets in the creamed layer does not occur usually, although it would be facilitated by creaming since the droplets are then much closer together. Stability to coalescence is determined by the nature of the interfacial film.

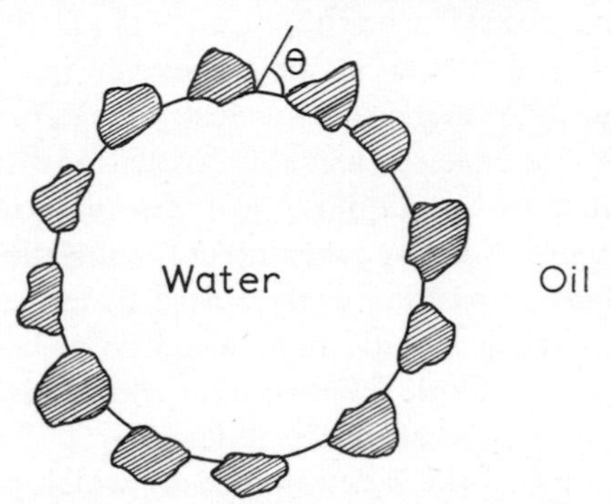

FIG. 27. Carbon black particles at the oil–water interface of a *W/O* emulsion droplet.

In the *breaking* of an emulsion the droplets coalesce and the two liquids separate out. This is important in certain technical processes in which undesirable spontaneous emulsification takes place. A variety of methods are used in the breaking of emulsions including heating, freezing of one of the liquid phases, addition of electrolytes, centrifugation with warming and the addition of substances that break down the stabilizing film. Emulsifying agents which produce *O/W* emulsions frequently break *W/O* emulsions and vice versa. A common example of the breaking of an emulsion is the manufacture of butter from cream. Intense agitation of cream causes the fat particles to coalesce into semi-solid lumps of butter which are then separated from the liquid phase.

FOAMS

A *foam* is defined as a dispersion of a gas (usually air) in a liquid. Most of the volume taken up by a foam is occupied by the gas bubbles which are separated by liquid films. Broadly speaking there are two extreme types of foam structure. In one case the bubbles are nearly spherical and separated by thick liquid films. Some of the liquid may drain away from the thick film under the action of gravity giving the other type in which the gas bubbles are polyhedral in shape, and the film thickness of the order 200–2000 Å.

Formation and Stability of Foams

Foam formation is achieved by simply shaking the gas and liquid together, or by bubbling the gas through a sintered glass disc into the liquid. When pure liquids are used the bubbles which form at the surface collapse rapidly. A stable foam is only produced when a third component, the *foaming agent* is present. There are a large number of surface active substances which are effective in the formation and stabilization of foams, the most important of which are the soaps, detergents, proteins and solid powders. In all cases the agent adsorbs at the interface and although the foaming ability depends on the surface activity of the agent, the stability of the foam is determined by the nature of the liquid film.

The stability of a foam depends on the rate of drainage of the liquid film, particularly in the early stages, and this in turn depends on such factors as the surface and bulk viscosities of the liquid film and on the elasticity of the film. All these properties determine how much drainage must occur before the film is so weak that it ruptures. In some cases the film may thin to about 50 Å but then ruptures by molecular forces. Considerable drainage occurs in foams formed by common soaps and detergents and consequently they are not very stable. As with emulsions, proteins and solid powders form quite rigid films and the foams owe their stability to the mechanical strength of these films.

Foams can be broken by a variety of methods depending on the properties of the particular system. The method used commercially involves spraying with organic compounds (e.g. ether, n-octanol) which, as a result of their greater surface activity, displace the foaming agent from the surface leaving a film without the requisite properties to resist rupture. For a similar reason foaming can be prevented by the addition of a suitable agent to the liquid phase: silicones are now widely used as foam inhibitors. The controlling of foam formation is of great importance in a number of technical processes.

Practical Aspects of Foams

The important industrial application of foaming in the froth flotation process for the separation of minerals has already been mentioned (Chapter 3). Another important type of foam is that used in fire fighting, particularly in oil fires. This foam is a dispersion of carbon dioxide gas in aqueous media containing a saponin of protein as foaming agent. The foam is fairly rigid and floats over the surface of the burning liquid as a "blanket" preventing the evolution of inflammable vapours.

Some common examples of foams found in the home include beaten egg-white (meringue), whipped cream, froth on beer, and those associated with cleaning processes involving soaps and detergents.

AEROSOLS

A colloidal dispersion of a solid or liquid in a gas is known as an *aerosol.* Mist and fogs are aerosols of liquid in gas whereas smoke, dust and fume contain solid at the disperse phase. However, this classification is not rigid since there are systems such as certain smokes which contain both liquid and solid particles as well as fogs which result from condensation of liquid on solid nuclei.

The undesirable aspects of most dusts, fogs and smokes are well known. Many industrial processes such as dry grinding, rock blasting and certain mining operations produce dusts which, if inhaled, may have harmful effects on the lungs. Siliceous dusts are particularly dangerous in this respect and lead to the disease silicosis. Dusts of certain inflammable substances, e.g. coal, flour, sulphur and sugar, under suitable conditions give explosive mixtures with air. For both cleanliness and health reasons the removal of smoke from the atmosphere is desirable, and various methods are now used to remove smoke and noxious fumes from flue gases in industrial plants. Washing with water is very effective since this removes both the suspended particles and such gases as sulphur dioxide. In the electrostatic precipitation method an electrical discharge causes the flue gases to ionize, and the ions are then adsorbed on the surface of the smoke particles. The charged particles are attracted to the electrodes on which they deposit. Smokes have, however, proved useful in war-time for camouflage purposes (smoke screens).

Aerosols are finding increasing applications in the home for removing objectionable smells, fly-killing, etc. Insecticidal aerosols are used for spraying fruit trees. Aerosol research has contributed to our understanding of various meteorological phenomena such as the formation of clouds and mists and the development of rain and snow in the atmosphere.

Preparation and Properties of Aerosols

As with lyophobic sols aerosols can be prepared by either dispersion or condensation methods. The simplest dispersion method involves forcing a liquid through a nozzle in an atomizer (like a scent spray) by means of compressed air or another gas. Aerosols produced in this way are usually polydisperse. In the condensation methods the first stage is the production of a supersaturated state in the gas phase. This is followed by nucleation and growth on the nuclei in a similar way to that already described for lyophobic sols. The supersaturation may be obtained by chemical reaction as on mixing gaseous hydrogen chloride and ammonia to form an ammonium chloride aerosol, or by super-cooling of a vapour by adiabatic expansion. When a high degree of supersaturation is attained nuclei form spontaneously, but for low supersaturations foreign nuclei must be present on which the vapour condenses. Air normally contains sufficient dust particles to act as nuclei for the condensation of water vapour at a relatively low degree of super-cooling forming a fog. These methods often lead to monodisperse aerosols.

Fogs and smokes appear at first sight to be much more stable than the common aqueous colloidal dispersions. The difference can be explained quite simply by considering the concentration of particles in the two cases. Aqueous sols have particle concentrations of the order of 10^{12} cm^{-3} while fogs and smokes usually contain 10^5–10^6 particles cm^{-3}. On the basis of the Smoluchowski theory of rapid flocculation a significant difference in stability would be expected although the viscosity of air is very much lower than that of water. Aqueous sols of similar particle size and concentration to those of the aerosols show the greater stability. Although the aerosol particles may be charged by collisions with naturally occurring ions in the air, the presence of a charge is usually of minor importance in stability to flocculation.

Evaporation of the aerosol droplets by an increase in temperature may cause a fog to disappear. This is the reason for the disappearance of natural fogs at sunrise. The droplets of town fogs may be covered with soot or tarry material which reduces the rate of evaporation and thus contributes to their stability.

GELS

The lyophobic and lyophilic sols described in Chapter 5 contain only a small amount of the disperse phase and normally flow quite easily. The term *gel* (or jelly) refers to a dispersion of a substance, usually lyophilic in character, in a liquid medium, which is comparatively rich in liquid and shows some measure of rigidity. Gels frequently contain only small amounts of the disperse phase, e.g. 1–3 per cent gelatine in table jellies, their rigidity being associated with the interaction between the particles which are linked together in some manner. The difference in flow properties between sols and gels may be taken as a means of classification although no clear-cut line of demarcation exists between them.

Formation of Gels

Gels may be formed from colloidal solutions in several different ways. Probably the most familiar example is the setting of a table jelly on cooling. Gelatine is soluble in hot water forming a molecular solution which on cooling to room temperature sets to a clear gel, if the concentration of gelatine is high enough. Setting or gelation occurs at a fairly definite temperature, the gelation temperature, and heating the gel to a few degrees above this temperature converts it back into a sol.

The addition of indifferent electrolytes to aqueous lyophobic sols causes flocculation and usually the particles settle out under gravity. However, there are cases for which gelation occurs upon the addition of salts, e.g. from aqueous aluminium and ferric hydroxide sols, provided the concentration of dispersed material is sufficiently high. The formation of gels upon flocculation is facilitated if the particles are linear although gelation does not necessarily occur when such particles are flocculated. For instance, gelatin and other protein sols show precipitation with high electrolyte concentrations.

Reduction of solubility by addition of a liquid in which the disperse phase is insoluble may also lead under appropriate conditions, to the formation of a gel, e.g. rapid mixing of a solution of polystyrene in benzene with petroleum ether.

Gels may also be formed by chemical reaction between concentrated solutions of reactants when one of the products is insoluble, e.g. a barium sulphate gel is obtained by mixing saturated solutions of barium thiocyanate and manganese sulphate. The insoluble material is thrown out of solution as a large number of colloidal particles which link together in chains to form a network structure throughout the system. The commercially important silica gel is formed by the reaction between hydrochloric acid and sodium silicate.

The Structure of Gels

Information on the structure of gels has been obtained from studies of the physical changes which occur during the sol–gel transformation and from observations on the gels using the ultramicroscope, electron microscope and X-ray techniques. The picture which emerges is that of a three-dimensional network structure of colloidal particles which immobilizes the liquid and imparts a certain degree of rigidity on the system.

There are three main types of gel which are classified according to their mechanical properties. Systems in the first group have a very stable network giving a rigid gel. Silica gel is a good example. When an acid is added to sodium silicate solution monosilicic acid is first liberated which immediately begins to polymerize to form a three-dimensional network with the following structure:

```
       |      |      |
       O      O      O
       |      |      |
 HO—Si—O—Si—O—Si—O—
       |      |      |
       O      O      O
       |      |      |
 HO—Si—O—Si—O—Si—O—
       |      |      |
      OH     OH     OH
```

Silica gel is rigid and stable since the silicic acid polymers are joined together through primary valencies. The water retained within the framework may be removed leaving a light amorphous solid with a large internal surface, which is used commercially as a drying agent.

The protein gels such as those of gelatin, and the pectin gels important in jam manufacture are members of the second group, which show elastic behaviour, i.e. they return to their original shape after removal of the applied stress. In these gels the colloidal units are linked together at a few points of contact by such forces as those of the coulombic type between opposite charges or by hydrogen bonds. The bonding is sufficient to confer rigidity on the structure although not strong enough to oppose deformation on application of a stress. These ideas are supported by the fact that the gel structure is reversibly produced or destroyed by changing the temperature.

In the third group the framework is so weak that it is destroyed (liquefies) by shaking but reforms on standing. This phenomenon of a reversible sol–gel transformation, known as *thixotropy*, is encountered in gels of ferric and aluminium hydroxides, and in certain types of paints.

REFERENCES

THE following are recommended for further reading:

ADAM, N. K., Water Repellency, *Endeavour*, **17**, 37 (1958).

ADAM, N. K. and STEVENSON, D. G., Detergent Action, *Endeavour*, **12**, 25 (1953).

ALEXANDER, A. E., *Surface Chemistry*, Longmans, London, 1951.

COSLETT, V. E., Electron Microscopy, *Endeavour*, **15**, 153 (1956).

DERJAGUIN, B. V., The Force between Molecules, *Scientific American*, 1960.

GRAY, G. W., Electrophoresis, *Scientific American*, 1951.

GRAY, G. W., The Ultracentrifuge, *Scientific American*, 1951.

JIRGENSONS, B. and STRAUMANIS, M. E., *A Short Textbook of Colloid Chemistry*, Pergamon Press, London, 1954.

KELLER, R. A., Gas Chromatography, *Scientific American*, 1961.

KITCHENER, J. A., *Ion Exchange Resins*, Methuen, London, 1957.

PANKHURST, K. G. A., *Detergent Solutions*, Royal Institute of Chemistry Monograph No. 5, 1953.

RIDEAL, E. K., Surface Chemistry in Relation to Biology, *Endeavour*, **4**, 83 (1945).

STEIN, W. H. and MOORE, S., Chromatography, *Scientific American*, 1951.

SVEDBERG, T., Molecular Sedimentation in the Ultracentrifuge, *Endeavour*, **6**, 89 (1947).

TAYLOR, R. J., *Surface Activity*, Unilever booklet, 1961.

WARD, A. G., *Colloids—Their Properties and Applications*, Blackie, London, 1948.

PROBLEMS

(WITH ANSWERS)

1. Calculate the height to which ethyl alcohol of density 0·7893 g cm^{-3} and surface tension 22·75 dyne cm^{-1} rises in a capillary tube of radius 0·2 mm, assuming that the walls of the tube are perfectly wetted by the liquid. (*Ans.:* 2·94 cm.)

2. The following surface tension data have been obtained at 25°C for an aqueous solution of hydrocinnamic acid:

Concentration (g/g of water)	Surface tension (dyne cm^{-1})
0·0035	56·0
0·0040	54·0
0·0045	52·0

Calculate the surface excess of the acid and the area occupied by each molecule in the surface when the bulk concentration is 0·0040 g/g of water. (*Ans.:* $6{\cdot}5 \times 10^{-10}$ mole cm^{-2}, 25·7 $Å^2$.)

3. The surface tensions of water and mercury are 72·0 and 476 dyne cm^{-1} respectively and the interfacial tension at the water–mercury interface is 369 dyne cm^{-1}. Calculate (*a*) the work of cohesion of water, (*b*) the work of adhesion between water and mercury, and (*c*) the initial spreading coefficient of water on mercury. (*Ans.:* (*a*) 144·0 dyne cm^{-1}, (*b*) 179·0 dyne cm^{-1}, (*c*) 35·0 dyne cm^{-1}.)

4. At 25°C a film of myristic acid spread on water exerts a surface pressure of 0·2 dyne cm^{-1} when each molecule occupies an area of 1800 $Å^2$. Calculate a value for the gas constant ***R*** assuming the film behaves as a two-dimensional ideal gas and compare the result with the accepted value for ***R***. (*Ans.:* $7{\cdot}28 \times 10^7$ erg deg^{-1} $mole^{-1}$.)

5. The contact angle for water on graphite at 25°C is 86°. Calculate the work of adhesion and the final spreading coefficient. The surface tension of water at 25°C is 72·0 dyne cm^{-1}. (*Ans.:* 77 dyne cm^{-1}, −67 dyne cm^{-1}.)

6. The following data have been obtained at 25°C for the adsorption of octadecanol from benzene solutions on to a carbon black of surface area 100 m^2 g^{-1}:

Equilibrium concentration (molar)	Amount adsorbed (micromole/g of adsorbent)
0·00791	16·63
0·03227	40·67
0·06340	52·50
0·09000	57·70
0·11500	61·00

Show that the data fit the Langmuir isotherm, and determine the amount of alcohol adsorbed and the area occupied by each molecule at saturation. (*Ans.:* 76 micromole g^{-1}, 218 $Å^2$.)

7. Calculate the "thickness of the double layer" associated with the particles of an aqueous dispersion at 25°C containing magnesium chloride in solution at a concentration of 10^{-5} M. The dielectric constant of water is 78·3 at 25°C. (*Ans.:* 550 Å.)

8. In a micro-electrophoresis experiment at 20°C, the large particles of an aqueous sol move through a distance of 700 μ in 20 sec under a potential gradient of 5 V cm^{-1}. Calculate the zeta potential. The dielectric constant and viscosity of water at 20°C are 80·1 and 0·01 poise respectively. (*Ans.:* 102 mV.)

9. Calculate the average displacement in 1 sec along a given axis produced by Brownian motion of particles of radius 100 Å in an aqueous sol at at 20°C. The viscosity of water at 20°C is 0·01 poise. (*Ans.:* $6{\cdot}6 \times 10^{-4}$ cm.)

10. By rapid flocculation the number of spherical particles (density 2·0 g cm^{-3}) in an aqueous sol at 20°C is reduced to half its initial value in 35 min. Calculate the apparent particle radius if the initial sol concentration is $8{\cdot}0 \times 10^{-6}$ g cm^{-3}. The viscosity of water at 20°C is 0·01 poise. (*Ans.:* 1000 Å.)

11. The critical concentration of potassium nitrate for the flocculation of a negative arsenious sulphide sol is $5{\cdot}0 \times 10^{-2}$ M. Estimate the probable flocculation values for magnesium and aluminium nitrates. (*Ans.:* $8{\cdot}1 \times 10^{-4}$ M, $7{\cdot}1 \times 10^{-5}$ M.)

12. The following viscosity data have been obtained for a sample of polystyrene in benzene:

η_{sp}	0·044	0·115	0·183	0·271
concentration (g/100 cm^3)	0·130	0·336	0·530	0·775

Calculate the molecular weight of the polystyrene assuming values of the constants $K = 2{\cdot}7 \times 10^{-4}$ and $\alpha = 0{\cdot}66$ for this system. (*Ans.:* 44,000.)

13. The osmotic pressure of an aqueous solution of egg albumin containing 10 g l^{-1} is $6{\cdot}03 \times 10^{-3}$ atm at 25°C. Estimate the molecular weight of egg albumin. (*Ans.:* 44,500.)

14. The following light scattering data were obtained for solutions of a polymer of concentration c (g cm^{-3}):

$c\ (\times 10^3)$	0·75	1·67	3·33	5·00
$(Hc/\tau) \times 10^6$	2·94	3·60	4·87	6·15

where τ is the turbidity and H is the Debye function containing the optical constants of the system. Determine the apparent molecular weight of the polymer. (*Ans.:* 425,600.)

INDEX